T0376404

GARDENING WITH
WINTER PLANTS

GARDENING WITH WINTER PLANTS

TONY HALL

Kew Publishing
Royal Botanic Gardens, Kew

Royal Botanic Gardens Kew

First published in 2023 by
Royal Botanic Gardens, Kew,
Richmond, Surrey, TW9 3AB, UK
www.kew.org

ISBN 978 1 84246 794 7

Distributed on behalf of the Royal Botanic Gardens, Kew in North America by the University of Chicago Press, 1427 East 60th St, Chicago, IL 60637, USA.

British Library Cataloguing in Publication Data
A catalogue record for this book is available from the British Library

Design and page layout: Kevin Knight
Project management: Georgina Hills
Copy-editing: Zia Allaway
Proofreading: Matthew Seal

Printed and bound in Italy by Printer Trento

For information or to purchase all Kew titles please visit shop.kew.org/kewbooksonline or email publishing@kew.org

Kew's mission is to understand and protect plants and fungi, for the wellbeing of people and the future of all life on Earth.

Kew receives approximately one third of its funding from Government through the Department for Environment, Food and Rural Affairs (Defra). All other funding needed to support Kew's vital work comes from members, foundations, donors and commercial activities, including book sales.

LEFT
Naturalised snowdrops

Contents

ABOVE
Frost-covered seed head of echinacea

Introduction

As temperatures fall and the frost and snow arrives, many gardeners hang up their tools and wait until spring. However, if you view winter as a dull and drab season with little to look at, think again. While many plants are resting at this time of year, others continue to flower throughout the colder months, and if you have yet to discover which plants perform in winter, you are missing out on the many joys this season has to offer.

The meteorological calendar, divides the seasons into four more or less equal periods made up of three months each, with winter running from 1 December until the end of February. But the seasons can also be defined by the astronomical calendar which is determined by the Earth's orbit around the sun. Using this system, the season starts on the winter solstice on 21 December, which is the shortest day of the year, and continues until around 20 March, the spring equinox, when day and night are of equal length.

Deciduous plants have usually lost their leaves as winter sets in, many having just put on spectacular autumn displays, when their foliage lights up the garden with fiery bronze, copper, orange, purple, red, and yellow colours before falling. This occurs when plants react to

Some of the range of autumn leaf colours hidden throughout the summer by green chlorophyll

a fall in temperatures by reducing the supply of fluids through their trunks, stems, and branches to their leaves. Chlorophyll, the green pigment in a leaf that helps plants to absorb energy from the sun during the process known as photosynthesis, also disappears, revealing other colours in the foliage that have been masked by this pigment throughout spring and summer. The chlorophyll has up until the end of summer been using the sun's rays to manufacture energy in the form of simple sugars. This is a continual process for half of each year, but as summer comes to an end, plants can sense this change and begin to reduce the amount of energy collected through their leaves, eventually closing off the vessels at the base of each leaf, an area known as the abscission layer. Without this supply of food, chlorophyll is no longer produced and so starts to break down, revealing fiery autumnal tones before trees shed all their leaves to start their winter rest.

With the stems now bare, many trees and shrubs reveal their skeletal forms, and in doing so show off their colourful trunks and stems, while fruits and seeds that appeared a few months earlier may also hang on into winter, offering additional seasonal interest.

There is also a wonderful range of plants that flower throughout winter, their blooms rivalling any that appear during other seasons of the year and lighting up the garden with their vivid colours and sweet fragrance.

Even the smallest of spaces can be brightened up with winter colour, and those without gardens can use their patios, courtyards and balconies to grow seasonal plants in containers, such as window boxes and hanging baskets.

ABOVE
The white stems of birches and vivid colours of cornus are revealed during the winter months

Early bulbs such as snowdrops look great planted in containers between the almost black leaves of *Ophiopogon planiscapus* 'Nigrescens', or with the deep purples, pinks, reds, yellows, and greens of hellebores, as well as ferns and heathers.

In larger spaces, swathes of blue and mauve *Crocus tommasinianus* naturalised in a lawn are a sheer delight in late winter and a great source of pollen and nectar for foraging bees taking advantage of any warmer winter days. These crocuses also work equally well in a pot, either planted alone or as a part of a mixed group.

During the summer months two of the most challenging spots for plants in a garden are areas in full sun, which can become very dry, and shady places that can also potentially be very dry, with low light levels. However, in winter, both of these conditions are generally less of a problem, as the soil is wetter and the shade cast from now leafless deciduous trees and shrubs much reduced. This extends the areas in a garden that can be planted in winter, and in the sunnier spots, you will also find insects out foraging among the flowers.

Weeds grow throughout the year, but less so in winter, thereby reducing the garden's maintenance needs and allowing you more time to look at and enjoy what you have, and possibly plan in a few extra plants to fill the gaps.

Many public gardens across the country have great winter displays to inspire you, including Anglesey Abbey in Cambridge, Sir Harold Hillier Gardens near Winchester, Kew Gardens in Richmond, Surrey, RHS Garden Wisley, also in Surrey, and Wakehurst in West Sussex. These and others display a great wealth of seasonal plants that can be grown in gardens large and small.

I personally enjoy the cold, crisp, sunny days of winter as much as similarly bright days in summer. You may need a few more clothes to keep the chill out, but that's to be expected, and walking among the colour and scent on offer at this time of year is a joy.

My own garden is one for all seasons, but I especially love creating a colourful garden for the colder months, when it's so uplifting to be outside. Research proves that outdoor green spaces, gardens, and gardening are all good for our wellbeing, including our mental health, so at a time of year when many of us are possibly at our lowest, visiting or creating a cheerful winter garden will definitely help to lift our spirits.

ABOVE
Naturalised *Crocus tommasinianus*, fully open in the winter sunshine

OVERLEAF
Wakehurst's winter garden lit up by the winter sun against a threatening sky

Berries and winter fruits

Berries and fruits offer a colourful element to any garden with winter interest, and many spring- and summer-flowering plants go on to produce them in autumn, some carrying them through into the colder months that follow.

These bright fruits are commonly known as berries, but, in fact, there is a distinction between the two. In botanical terms, a berry is defined as a simple fruit derived from a single ovary of an individual flower. So, strangely, a banana is technically a berry, as is a grape and a tomato. However, blackberries, raspberries, and strawberries are not true berries but 'aggregate fruits', which are produced from flowers that have multiple ovaries. Even holly berries are technically not berries but drupes.

It's all very confusing, so to make things easier, I refer to them all in the book as simply berries or fruits.

The most well-known winter berries are those of the holly (*Ilex*). As well as being associated with winter, they have long had a connection to Christmas, and are frequently represented on cards and wrapping paper, their clusters of shiny red berries often dusted with frost or snow. Holly branches are also often used in wreaths and to decorate the home during the seasonal celebrations. Although hollies are generally featured with red berries, the fruit comes in various shades of orange, yellow, purple–black, and even white (*Ilex glabra* f. *leucocarpa*).

Only the female plants produce fruits, but a male tree is also needed nearby to guarantee a crop of berries each year. As well as the colourful fruits, evergreen Ilex varieties with variegated leaves also offer year-round interest.

Callicarpa bodinieri, commonly known as the beauty-berry, is another plant with striking winter interest. The violet–purple fruits are produced in late autumn when they contrast well with the pink and purple-tinted leaves. After the foliage has fallen, the attractive berries, which appear in clusters along the bare branches, become the stars of the show and often last well into winter. The cultivar *C. bodinieri* var. *giraldii* 'Profusion' is one of the best, its large clusters of violet–purple berries on arching stems decorating the garden and also making a long-lasting addition to winter flower arrangements for the home.

Many *Cotoneaster* species are evergreen, offering more than one season of interest, and they all bear clusters of berries in autumn and winter. Most are dark red, but orange- and yellow-berried cultivars are also available. There are many different forms, too, from small trees such as *Cotoneaster lacteus*, to low-growing, mound-forming shrubs, including *Cotoneaster horizontalis*, which can also be trained against a wall – its branches are

Winter berries of holly have a long association with Christmas, here with variegated foliage (*Ilex aquifolium* 'Northern Lights')

festooned with red berries from late autumn into winter.

Mistletoe (*Viscum album*) is another plant that has a long association with winter and Christmas. Early Celtic priests, known as druids, from around the 1st century CE thought that mistletoe had magical and healing properties and, among its many powers, it could help to restore fertility. More recently, it was thought that kissing under mistletoe would lead to marriage, and to refuse a kiss was bad luck.

Mistletoe is an epiphyte, which means it grows on a host plant, and dioecious, producing male and female flowers on separate plants – the white translucent berries only appear on the females. It is most often brought into the garden by birds, which after eating the flesh of the berry wipe their beaks on the branches of trees, depositing the seeds in the process. The seeds can also be in birds' poo and, likewise, this can end up on tree stems and germinate to produce these unusual evergreen plants.

If you want to introduce mistletoe into your garden, you will need to purchase some fresh berries and rub or push the seeds into the fissures of the bark along a horizontal branch. You are more likely to be successful if your host tree is mature and an apple, hawthorn, lime, or poplar. Also be patient, as it may take a few years before you see any results.

Nandina domestica, the heavenly bamboo, is mainly grown for its evergreen foliage, which changes colour throughout the year. But in autumn it also produces green berries that ripen to bright red and persist well into the winter months, probably because birds and animals do

not find them palatable. There is also a creamy-white berried form called 'Alba'.

Firethorns (*Pyracantha*) are tough fruiting shrubs, and the cultivar 'Orange Glow' offers an abundance of bright orange berries in autumn and winter that stand out well against the glossy evergreen foliage. Other cultivars produce red or yellow berries. As well as being very attractive, firethorns are also a great source of food for birds, which flock to the plants during the winter months. The downside to this, of course, is that in many cases the berries are soon eaten, although the yellow-berried cultivars are the least attractive to birds and last the longest.

Many rose species produce excellent autumn fruits known as hips, which endure through winter, when their appearance is enhanced by frost or a dusting of snow.

Hips come in a variety of sizes and colours, and include the dark purple to almost black fruits of the Scotch rose, *Rosa spinosissima* and the clusters of bright red flagon-shaped fruits of *Rosa moyesii*, which can be up to 5 cm/2 in in length. Other good choices include the rambling *Rosa filipes* 'Kiftsgate', which produces much smaller hips but in great profusion – just make sure you have space for this large plant, which can grow to a height and spread of 10 x 6 m/30 x 20 ft. The hips of *Rosa rugosa* and its cultivars are among the biggest, with long sepals at the top that look like five twisted tentacles.

Skimmias are great little shrubs for year-round interest. They are dioecious (see above) and only the female plants bear berries throughout the winter,

ABOVE
The translucent berries produced on the female plants of mistletoe

though male plants are also needed to ensure they fruit. However, there are also hermaphrodite forms such as *Skimmia japonica* subsp. *reevesiana* that will produce berries without the need for a companion plant. Their long-lasting berries are usually red, but both black- and white-berried cultivars are also available. Skimmias also make good small shrubs for containers and some have fragrant flowers in spring too.

The snowberry (*Symphoricarpos albus*) is an apt name for this white-berried deciduous shrub, which is very happy growing in sun or shade and tolerant of most soil conditions. The berries stand out well among the foliage, but they become even more showy once the leaves have fallen.

Viburnum nudum 'Pink Beauty' and 'Brandywine' are two cultivars with clusters of berries that change colour as they mature, from white to pink and, finally, either dark blue or black. Their berries are produced late in autumn as the foliage takes on its autumnal hues and they remain on the plant after the leaves have fallen.

There are many other plants that produce attractive fruits and berries during winter, including perennials such as *Iris foetidissima*, which produces bright, shiny orange berries. It also tolerates a shady position in the garden.

TOP
A touch of frost adds extra interest to many winter fruits, here on *Rosa rugosa*

ABOVE
A dusting of snow on *Cotoneaster lacteus*

Bulbs

Hardy species of bulbs and their cultivars are a great addition to any winter garden, and top of my list are the snowdrops (*Galanthus*). These classic winter-flowering plants include many different species and varieties that offer a wide range of flower sizes and heights. Most are white with green markings although some have yellow colouring, and there are also double-flowered and even scented flowers to choose from.

Other early-flowering bulbs include crocuses, cyclamen, and winter aconites. While not strictly bulbs, these corms and tuberous perennials are usually grouped together as such, and will deliver a wealth of colourful flowers to the winter garden, either when planted as individual species, or mixed together in a bed or border. They look particularly good beneath deciduous trees and shrubs, where they grow naturally in the wild, taking advantage of the extra light beneath the canopies before the plants overhead come into leaf and shade them out.

Some daffodils flower early in the year, too, including 'February Gold', a popular cultivar that has been around for a long time and comes back reliably year after year. Another early-flowering species is the British native, *Narcissus pseudonarcissus*, known as the wild daffodil, or Lent lily. The flowers comprise white or pale yellow petals (tepals), and a darker yellow trumpet (corona), and their appearance in February is a sure sign that spring is just around the corner.

The dwarf *Iris reticulata* and its many different colour forms will brighten up the front of a border and also look great planted in containers, while the slightly taller *Iris unguicularis* is more suited to growing in beds and borders.

Blue-flowering bulbs such as glory-of-the-snow, *Scilla forbesii*, and the Siberian squill, *Scilla siberica*, will add a cool note to your winter colour scheme. These come into flower right on the cusp of the seasons as winter moves into spring and both will naturalise well.

Most winter-flowering bulbs do well in containers, bringing colour and interest to a patio or planter on a windowsill, where they can be enjoyed from the warmth indoors. But for sheer flower power, grow them in large drifts for a breath-taking display, or allow them to naturalise in short grass – purple–blue carpets of the early crocus, *Crocus tommasinianus*, for example, quickly spread to produce a stunning feature, while snowdrops (*Galanthus nivalis*) do equally well, multiplying in their hundreds or thousands. Drifts of cheery yellow daffodils will also naturalise in a lawn, increasing each year,

Both snowdrops and aconites
naturalise and work well together

though they do just as well in beds and borders, where buttercup-like winter aconites also thrive.

To achieve a naturalistic planting, just scatter handfuls of bulbs and plant them where they land – if two or three land together, simply plant them that way. They will then look totally natural, an effect you won't achieve if you try to place them randomly, however hard you try.

Most spring bulbs are bought and planted in the autumn, and are generally planted at two or three times the depth of the bulb. However, planting bulbs in the green (in leaf, just after the flowers have faded in spring) can be a better way to establish some species, most notably snowdrops, than planting dry stored bulbs in autumn. This allows them to grow more quickly as their roots are still active, and the leaves are still able to feed them. Planting dry bulbs will be okay, too, but they may take a bit longer to establish. Bulbs planted in the green should always be set at the same depth in the soil as they were when originally lifted. Check where the stems change from white to green and plant so that only the latter part is showing above the surface.

Use the same principle when you divide overcrowded bulbs. Lift, divide, and replant them as soon as possible after flowering to give them a chance to re-establish before they go dormant in summer, which will give them a head start for the following season.

Cutting the grass where bulbs have naturalised should be left until after all their foliage has died back. This allows the leaves to continue to feed the bulbs and store more energy, producing stronger plants for their next flowering season.

Another great way to enjoy winter bulbs is to grow some in pots and bring them into the house when they are in flower. An obvious choice is daffodils, but crocuses and dwarf irises work equally well.

Or you may wish to grow tender bulbs indoors such as amaryllis, also known as *Hippeastrum*. These elegant plants produce tall, stout stems, topped with large trumpet-shaped flowers that bloom about eight weeks after planting, allowing you to time them to appear in succession throughout the winter.

Scented hyacinths bulbs can be forced to flower indoors for Christmas by starting them about three months beforehand at the end of September. Plant the bulbs (buy those marked 'prepared') in pots with a 5 cm/2 in layer of compost below them and then top up with more compost so that just the pointed tips are showing. Place the pots somewhere warm and dark and keep the compost moist. Once shoots start to appear, bring them out into the light – this usually takes about eight to ten weeks. They should then start to flower after two to three weeks.

OPPOSITE
Wild daffodils, *Narcissus pseudonarcissus* covering the mound in the woodland garden at Kew

ABOVE
Potted snowdrops work well with dark grass-like leaves of *Ophiopogon*

Foliage

Evergreen foliage plants offer leaves in a broad spectrum of colours to decorate the winter garden, and many have the added bonus of flowers, berries or seedheads that can provide interest at other times of year too. They are particularly useful where space is limited and you need multi-purpose plants that earn their keep.

The foliage of these trees, shrubs, and perennials comes in a wide range of shapes, sizes, and colours, each enhanced in winter by a covering of sparkling frost or dusting of snow.

The larger evergreens can perform as a colourful screen to block out eyesores such as old buildings or to mask a neighbour's window and offer you more privacy. Alternatively, some can be clipped to create a hedge, or used as a backdrop to show off other plants, contrasting with and highlighting those in front. They can also make good specimens in their own right and, in the case of the many golden and variegated plants, brighten up an otherwise dark area of the garden.

One of the most popular evergreen trees with winter interest is, of course, the holly (*Ilex*). Not all hollies are garden-worthy plants, but many species and cultivars are ideal and offer a host of leaf colours and forms to choose from. They also have appealing clusters of shiny berries in a variety of different colours, most plants holding on to them in winter and helping to brighten up what can be the drabbest months of the year.

Their thick, leathery, waxy leaves range from plain dark green to those with white, cream, or yellow variegations. Sometimes the variegation is confined to the leaf margins, or splashes of colour may be in the centre of the foliage, while in others, the pattern is completely random. Holly foliage may also have spiny or smooth margins, or prickles covering the whole leaf surface – these are known as hedgehog hollies. The plants themselves also vary in size, some forming small shrubs, others growing into large trees, with a selection to suit almost any garden.

Aside from the hollies, there are many other beautiful evergreens suitable for a winter garden. For example, conifers are perfect for medium-sized and large gardens, while the dwarf forms, such as cultivars of *Pinus mugo*, will fit into the tiniest of spaces. Small shrubs including the skimmias and sweet box (*Sarcococca*) are also ideal for small gardens and do well in containers, too.

Many ground-cover evergreen plants, including perennials such as heucheras, ferns, periwinkles (*Vinca*) and lungworts (*Pulmonaria*), thrive under deciduous trees and shrubs once they have dropped their leaves, and also tolerate the dappled light under their canopies for the rest of the year.

When choosing climbers for winter interest, consider an ivy (*Hedera* species). A vast range of species and cultivars offer a wealth of leaf shapes and colours, as well as plant sizes. There are also winter-flowering clematis that deliver foliage colour, flowers, and attractive fluffy seedheads.

Most evergreen foliage also lasts well when cut and used in a vase indoors, either on its own or mixed with other leaves and winter-flowering scented plants. Guaranteed to brighten up any windowsill or table, it offers yet another way to enjoy winter's natural bounty.

ABOVE LEFT
Golden-yellow variegated leaves of *Elaeagnus pungens* 'Maculata'

ABOVE RIGHT
The frosted leaves of *Heuchera* create patterns along their wavy margins

Fragrance and wildlife

One of the best features of a winter garden is the heady scent that many flowering plants produce. Shrubs such as wintersweet (*Chimonanthus*) and witch hazels (*Hamamelis*) have sensational, intoxicating fragrances that will draw you to the plants from quite a distance, and, amazingly, the scent is even produced when the blooms are covered in frost.

So why are these winter flowers among the most fragrant of all, appearing when there are fewer plants in bloom than in any other season?

While these flower fragrances are attractive to us humans, for the plant, there is a more important reason to go to all the effort of producing strongly scented, pollen-rich blooms when many other plants are dormant.

A flowering plant's main purpose is to produce flowers that will be pollinated, ensuring that they will eventually produce seed and then that seed will germinate. In this way, they guarantee the continuation of the next generation of plants, so something is needed to attract the attention of insects that will move pollen from the male to the female parts of their flowers to fertilise them.

There are, of course, winter-flowering trees and shrubs such as willows (*Salix*) and the silk tassel bush (*Garrya*) that bear catkins and tiny flowers that are wind pollinated and don't need to lure pollinators. But these are not fragrant (at least not to us), and although they do attract some pollinating insects, their services are not required to help with fertilisation, since the fine dust-like pollen is easily carried between their flowers by the lightest of breezes.

With fewer pollinating insects around during the winter months, those plants that need them have to work hard to entice those that are out and about on milder days. Emitting a strong fragrance helps to pull them in. Flowering in winter can also be advantageous for the plant, since there is less competition from other blooms and therefore they tend to get the undivided attention of pollinators and reward them for visiting with high-sugar and high-protein food.

Flowers that are pollinated by bees and butterflies (yes, even the odd butterfly can be seen on some of the mildest winter days) usually release their powerful scents during the day because the nights are generally far too cold for the insects, while many summer-flowering plants emit their scent after dusk as well to attract night-flying moths.

There are a surprisingly wide variety of plants that

ABOVE LEFT
Sweetly scented flowers of *Chimonanthus praecox*

ABOVE RIGHT
Lacking scent, wind-pollinated flowers/catkins are an attractive winter feature nonetheless

produce fragrant flowers at this time of year, from trees and shrubs to perennials and bulbs. The golden crocus (*Crocus chrysanthus*), for example, is highly scented and produces large amounts of pollen and nectar for early flying insects, particularly the queens of the buff-tailed bumblebee (*Bombus terrestris*), which are often out foraging after their winter hibernation. For them, it is the beginning of a new season, when they are preparing to produce a new colony of bees and stocking up on food supplies to take back to a developing nest.

In the warmer parts of Britain, some bumblebee nests will have over-wintered and the queens and workers will be active throughout the winter on mild days. You will even find some out foraging in taller shrubs such as mahonia when there is snow lying on the ground.

The fragrance produced by the witch hazels (*Hamamelis*) also attracts small flies and moths such as the satellite moth, *Eupsilia transversa*, which serve as winter pollinators.

So keep an eye out around your winter-flowering plants – I'm sure you will be surprised at just how many insects are active on the milder days of winter.

ABOVE LEFT
An early foraging bumblebee enjoying the nectar of mahonia

ABOVE RIGHT
On the mildest winter days even surprise foragers like this overwintering red admiral butterfly will venture out

Stems

Deciduous plants with brightly coloured and textured stems, branches, and trunks provide a beautiful display in winter, after their leaves have fallen.

Each year most trees shed some of their bark, but many also do it in a way that makes an attractive feature. Shedding bark is thought to be a natural process to remove disease and pollution that may have accumulated on the old bark. This was one of the reasons *Platanus* x *hispanica*, commonly known as the London plane, was planted by the Victorians – its ability to shed its bark each year removed the heavy pollution that London and other cities suffered during that period. This self-cleansing process also results in a beautiful stem feature, as the small plates of bark that fall away reveal different shades beneath to produce a multi-coloured effect, similar to the bark of the lacebark pine, *Pinus bungeana*.

Many birch species also produce decorative stems as the old bark peels away and uncovers a contrasting lighter-coloured bark beneath. The paperbark maple, *Acer griseum*, and the Tibetan cherry, *Prunus serrula*, both hang on to much of their peeling bark, often for more than one season, and make very appealing trees for a winter display.

Shedding their leaves in autumn has many benefits for deciduous plants, helping to protect them against the inclement weather conditions in winter. Those that retain a leafy canopy during the colder months may suffer more during seasonal storms when wind can tear the foliage or take hold of it and uproot the plant. Ice and snow can also cause damage when the cells in the leaves become frozen and are unable to photosynthesise. (Most evergreen leaves have thick outer walls, known as *cuticles*, to help prevent this.)

Discarding leaves that have been damaged by insect attacks, or viruses and diseases, also helps to protect the plants. The fallen leaves are not wasted, however, as the nutrients they hold are then released back into the soil during the decaying process, forming fertilising organic matter from which the plant recoups those it needs.

The benefit of leaf-fall to the gardener is that during this dormant period, many plants reveal an array of brightly coloured stems, from the powdery whites of ornamental brambles (*Rubus*), though to the deepest purples of the dogwood *Cornus alba* 'Kesselringii'.

The dogwoods are probably the most striking of all plants with winter stem interest, particularly when they are planted en masse. The lime-greens, yellows, and reds light up the garden in winter, some offering a mix of shades, such as 'Midwinter Fire', which, as its name suggests, creates a living bonfire, whose fiery red and yellow stems deliver a blaze of colour on cold grey days. I often cut different coloured stems from a range of

ABOVE
Colourful winter stems of cornus, white-stemmed bramble, and willow

TOP RIGHT
The stunning bark of *Prunus serrula* is often referred to as looking like polished mahogany

ABOVE RIGHT
Acer griseum, with its distinctive peeling bark

OPPOSITE
The striking winter stems of *Cornus* 'Midwinter Fire'

cultivars, standing them in a tall vase in the house, which I think looks amazing.

The stark white, creamy-pink, and coppery-brown trunks of the many different birch cultivars add to the winter display, their peeling bark revealing different tones as the old is replaced by the new.

The Tibetan cherry, *Prunus serrula*, is another good choice, its deep red stems shining like polished mahogany in the winter sun, while *Acer griseum*, the paperbark maple, bears rich cinnamon-coloured, papery, peeling bark, as its common name suggests.

Even in a small garden there is room for plants with colourful stems. A single multi-stemmed tree will look beautiful with just a couple of dogwoods (*Cornus*), or an underplanting of grasses, sedges, ferns, and winter-flowering bulbs such as snowdrops and winter aconites.

There are so many colourful stem combinations that will bring joy in the depths of winter, especially if they can be viewed from the warmth indoors when it is too cold to venture out.

Winter containers

Growing plants in containers, whatever the time of year, will bring colour and interest to the garden, particularly in small spaces, but in winter pots of flowers and foliage are especially useful, helping to lift the spirits while benefiting foraging insects out looking for nectar and pollen.

Before deciding what you want to plant in your winter display, first consider the type of container you will need. When buying a clay or terracotta pot, choose one labelled frost-proof, rather than frost-resistant, which should ensure that it makes it through the winter without cracking when temperatures fall.

Those made from fibreglass, metal, or recycled plastic should also be able to withstand frost, while most wooden planters will last a few winters and look more naturalistic. Choose a container from the huge range of different styles and colours on offer that complements your garden design and plant choices. Also ensure that it has drainage holes, or that you are able to make some, in the bottom to prevent the compost becoming waterlogged, rotting the plants' roots and, in freezing conditions, potentially damaging the container.

Choosing plants is a matter of personal taste, but it is good to have a theme, combining those in complementary or contrasting colours and forms, and perhaps adding a few scented varieties to create a stunning, long-lasting display. Options could include foliage plants such as *Euonymus fortunei* 'Emerald Gaiety', grasses, and heucheras, or ivies, sedges and ferns in shadier spots, together with flowers or berries to add a touch of colour. Winter-flowering plants such as cyclamen, heathers, hellebores, polyanthus, violas, and wallflowers will introduce interest, while gaultherias and skimmias offer clusters of bright red berries. Alternatively, try the evergreen *Skimmia* 'Rubella', a male form that does not produce berries but delivers clusters of tight pinkish-red buds throughout winter, opening to fragrant white flowers in early spring. Other good choices for a pot or small container are winter aconites, crocuses, snowdrops and other winter-flowering bulbs.

In large containers, you can add height with a variegated cordyline or the coloured stems of dogwoods, which I think go well with hellebores and snowdrops, plus the black leaves of *Ophiopogon planiscapus* 'Nigrescens'.

If you are planting a large container, put it in place before filling it, since it will be much more difficult to move to its final position afterwards. Placing your pots and containers on small plant pot feet or bricks to raise them off the ground not only allows water to drain

Winter container showing the attractiveness of just using mixed foliage plants

Snowdrops work well in a mixed planting in a container, but equally well planted as a group on their own

through them more easily but also prevents contact with any damaging frozen ground conditions.

When planting, start by placing some crocks in the bottom of the container. These could be pieces of broken clay pot, or perhaps some largish pebbles that would also add weight to the container, making it less likely to blow over. Then half fill it with good-quality compost. The type of compost you need will depend on the plants you are growing. If you are using shrubs and perennials, consider using a peat-free soil-based compost, which is better suited to semi-permanent plantings, while a general-purpose compost is much lighter in weight and will be fine for hanging baskets, smaller containers and pots, particularly if they are planted with short-term displays that will be re-used and planted in the garden once they finish flowering.

It's a good idea to have a dry run when placing your plants. Set them in the container on top of the compost while they are still in their individual pots to see how they fit and look together, moving them around until you are happy with the combination. Plants can be planted closely and almost squeezed in, because during the winter months they will grow very little, if at all, and so your plants need to make an instant impact.

Then it's just a matter of removing them from their pots, planting them and adding more compost around the rootballs, firming gently as you go. Make sure that your plants are at the same depth as they were in their original pots – do not bury the leaves and stems – and leave a gap of about 5 cm/2 in between the compost and rim to allow space for watering.

Once planted up, check the container regularly to make sure the compost does not dry out, particularly during mild periods.

Hanging baskets also make great containers for winter displays. Use smaller plants as the root space is much more limited. Good choices for baskets include miniature daffodils such as *Narcissus* 'Tête-à-tête,' *Iris reticulata*, and snowdrops.

Bugle (*Ajuga*), cyclamen, small ferns, heathers, pansies, primroses, sedges, and variegated ivies are other suitable plants for baskets; garden centres and online nurseries may offer an even wider selection. As with the larger containers, make sure the compost does not dry out and bear in mind that because they are smaller, you may need to water them more frequently. Once a week is probably sufficient but keep an eye on them because too much water is as big a problem as letting them dry out.

With the arrival of spring, some plants in your containers will come back into growth, at which point you can give any that threaten to outgrow their space a second life in a bed in the garden, where you can continue to enjoy them.

Bulbs

The plants in this section cover bulbs, corms, rhizomes, and tubers. Within each plant profile it will state which type applies.

CROCUS CHRYSANTHUS

Golden crocus

IRIDACEAE

The species *Crocus chrysanthus* has golden-yellow, goblet-shaped flowers, with bright yellow stamens and a rich orange stigma. Each corm can produce up to four individual flowers, although most will bear just one or two. The narrow leaves are dark green with a central silver stripe.

The cultivar **'E. P. Bowles'** has the same golden flowers but with darker bronze–purple stripes and feathering on the outer surface of the petals, which makes it attractive even when the flowers are not open.

'Cream Beauty' is a deep cream colour, with a golden-yellow centre.

'Romance' has pale buttery-yellow blooms, deeper yellow on the inside, with a dark blotch at the base of the petals. It will naturalise but tends to spread in small clumps.

ASPECT: Full sun, partial shade
FLOWERING: February – March
HARDINESS: –10 to –15°C (14 to 5°F)

5–10 cm (2–4 in)

10 cm (4 in)

CROCUS TOMMASINIANUS

Early crocus

IRIDACEAE

As the common name suggests, *Crocus tommasinianus* is one of the earliest crocuses to bloom, producing lilac to deep purple goblet-shaped flowers. During wet periods, dull days, and at night the flowers close completely, but open fully in bright sun, putting on an outstanding show.

This crocus is an excellent choice for naturalising in grass, where it will spread relatively quickly to produce a carpet of flowers that are very attractive to insects, particularly bees, early in the year.

The best time to plant the corms is in the autumn, at a depth of roughly twice the height of the corm.

Grass should be left uncut until after all the leaves have died down (around six weeks) to allow them to feed the bulbs and produce a good show of flowers the following season.

The cultivars **'Ruby Giant'**, a rich red–purple, and **'Whitewell Purple'**, which is purple with silvery tones on the inside, will also naturalise well.

ASPECT: Full sun
FLOWERING: February – March
HARDINESS: –10 to –15°C (14 to 5°F)

5–10 cm (2–4 in)

5 cm (2 in)

CYCLAMEN COUM

Eastern cyclamen

PRIMULACEAE

Cyclamen coum is a tuberous perennial with long-lasting leaves that are almost as attractive as the flowers.

The heart-shaped foliage appears in late autumn, each leaf decorated with a unique marbled pattern; they generally have a dark centre surrounded by silvery-green, with a darker green edge, but they can be very variable.

The flowers then appear around Christmastime in late December. Each flower, with its upswept petals, is held on a single stem above the foliage and colours range from shades of pink to deep purple.

The flowers of the white form, *Cyclamen coum* subsp. *coum* f. *pallidum* 'Album', have a purple blotch that forms a ring at the base of the petals.

Crocus tommasinianus

Flowers in shades of pinks and whites contrast well with the ground-covering marbled, heart-shaped leaves

Cyclamen coum

Cyclamen do best in humus-rich soils, where they will slowly spread and naturalise.

ASPECT: Best in partial shade
FLOWERING: December – March
HARDINESS: –10 to –15°C (14 to 5°F)

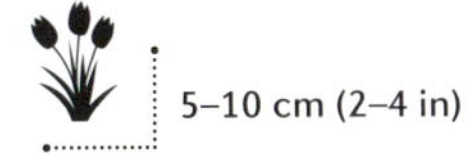

CYCLAMEN HEDERIFOLIUM

Ivy-leaved cyclamen

PRIMULACEAE

Although this is an autumn-flowering tuberous perennial and the fragrant flowers are already over before the start of winter (although, occasionally, a few will bloom in early winter), the leaves of this cyclamen are very attractive and make a carpet of ground cover throughout the winter and spring, often looking good up until the end of April.

The heart- or ivy-shaped leaves are variegated, each displaying a slightly different mottled grey–green pattern. Winter-flowering bulbs such as snowdrops and winter aconites contrast well against the darker foliage of this cyclamen.

ASPECT: Best in partial shade
FOLIAGE INTEREST: November – April
HARDINESS: –10 to –15°C (14 to 5°F)

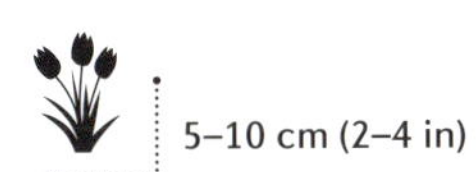

ERANTHIS HYEMALIS

Winter aconite

RANUNCULACEAE

Eranthis hyemalis is a species of plant in the same family as buttercups and the similarity is plain to see, its bright yellow shiny petals forming small cup-shaped flowers that glisten in the winter sunshine. These tuberous perennials occur naturally in woodlands and are ideal for parts of the garden that are in semi-shade, but they will also tolerate fairly deep shade, as well as full sun. Each individual flower is produced on a single stem and is set off by a characteristic ruff of divided leafy bracts.

Winter aconites can be a little difficult to establish and are best planted in the green (in leaf), but once they get going, they will multiply quickly and naturalise. They look particularly good growing under deciduous trees among snowdrops and hellebores.

At the end of their flowering season, they can be left to die back naturally. This will allow them to produce and spread their seed.

ASPECT: Full sun, to full shade. best in partial shade
FLOWERING: January – February
HARDINESS: –10 to –15°C (14 to 5°F)

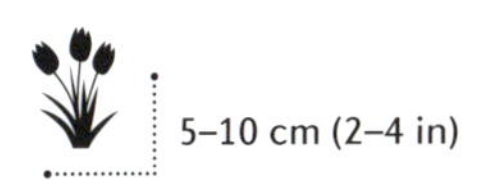

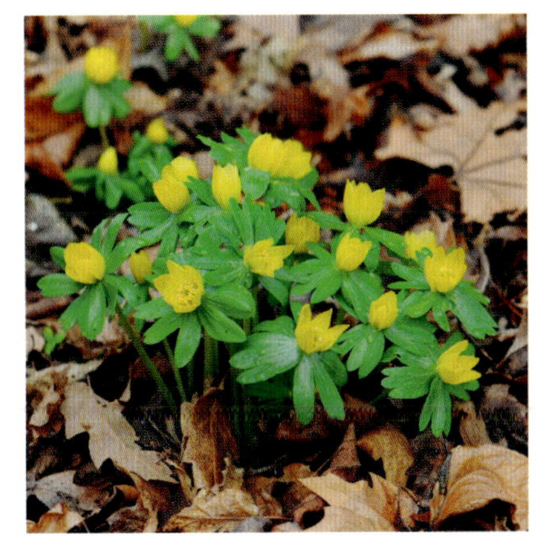

Eranthis hyemalis

GALANTHUS

Snowdrops

AMARYLLIDACEAE

A winter garden would not be complete without snowdrops. Flowering in the depths of winter, their dainty nodding white flowerheads brighten the greyest of days and appear when little else is in bloom, offering a reminder that spring is not too far away.

They are also known as Candlemas bells, as the flowers have a long association with the Christian festival of Candlemas at the beginning of February.

The snowdrop is one of many plants that have been used to record flowering times (phenology). Kew has been keeping a record of snowdrops since the 1950s, when they most frequently flowered in late February, but in more recent years, they have been flowering much earlier in January and, occasionally, even earlier in mild winters, providing an indicator of our changing climate.

There are approximately 20 species of snowdrops, and from three of the most common species, *Galanthus elwesii*, *Galanthus nivalis* and *Galanthus plicatus*, hundreds of varieties have been produced.

The flowers have six petals (tepals), three large ones in the outer whorl, and three smaller ones in the inner whorl. In most snowdrops, the characteristic green markings are on the inner tepals, but both the markings and petals shapes can vary greatly in the huge range of cultivars, with some displaying green patterns on all petals. There are also yellow cultivars, double-flowered varieties

Forming carpets of nodding white flowers, naturalised snowdrops are one of the earliest winter flowers to put on a show

Galanthus nivalis

and a cultivar of *Galanthus elwesii* named 'Godfrey Owen' that has twelve petals, six outer and six inner.

People who collect snowdrops, or snowdrop lovers, are known as galanthophiles, and they will pay high prices for a single bulb of something new. In 2022 the record price for one bulb was £1,850.

Galanthus nivalis is one of the most common snowdrops in cultivation around the world and it is also one of the easiest to grow for naturalising.

GALANTHUS NIVALIS
– Common snowdrop

The common snowdrop is surely a plant that needs little description – when most people think of winter flowers, I am sure that this is one of the top five that comes to mind. In fact, it naturalises so well in the UK (as well as in many other counties), we could be forgiven for thinking that it is a native, as it's seen growing in many mature woodlands, and along countryside hedgerows. But, in fact, the common snowdrop is actually native to much of southern Europe.

The small, white, nodding, bell-shaped flowers are borne singly, with three outer petals (perianth segments) that are usually pure white, while the inner three, which are much smaller, have decorative green markings most commonly in the shape of a U or V. Each flower hangs from a thin arching pedicel (stalk) within a sheath-like bract at the top of an upright stem, surrounded by two or occasionally three glaucous green leaves.

In the wild, this species is mostly found growing in deciduous woodland in humus-rich soils, and this is the situation that suits it best.

'Anglesey Abbey' was originally found at Anglesey Abbey, hence the name, and has an inverted V marking on the tips of the inner petals. Height: 15 cm/6 in, flowers from mid-January.

'Lady Elphinstone' is a double yellow form with yellow markings on the insides of the inner petals. Height: 10–15 cm/4–6 in, flowers from February.

'Magnet' has a distinctive long pedicel, which allows it to sway around in the breeze. It naturalises well from offsets (small bulbs). Height: 20 cm/8 in, flowers from January.

'Viridapice' has green markings on both inner and outer petals, with V-shaped patterns on the inner ones

Galanthus nivalis 'Anglesey Abbey'

Galanthus nivalis 'Magnet'

Galanthus nivalis 'Viridapice'

Galanthus nivalis Sandersii Group

Galanthus nivalis 'Lady Elphinstone'

Galanthus elwesii 'Esther Merton'

Galanthus elwesii 'Godfrey Owen'

Galanthus elwesii 'Grumpy'

Galanthus elwesii 'Margaret Biddulph'

Galanthus plicatus 'Diggory'

Galanthus plicatus 'Greenfinch'

Galanthus elwesii 'Cinderdine'

Galanthus plicatus 'Trymlet'

and green blotches on the tips of the outer ones. Height: 15 cm/6 in, flowers from February.

Sandersii Group covers the forms of *Galanthus nivalis* with yellow rather than green markings, and includes 'Ray Cobb' and 'Savill Gold'. Height: 15 cm/6 in, flowers from February.

GALANTHUS ELWESII
– Greater snowdrop

As its common name suggests, the greater snowdrop is around twice the size of *Galanthus nivalis*.

It has broad grey–green leaves and large flowers, each of which has two green markings on the inner three segments that are often joined together and appear as one.

'Cinderdine' has long, pure white textured outer segments, with a green marking on the tips of the inner segments. Height: 15 cm/6 in, flowers from February.

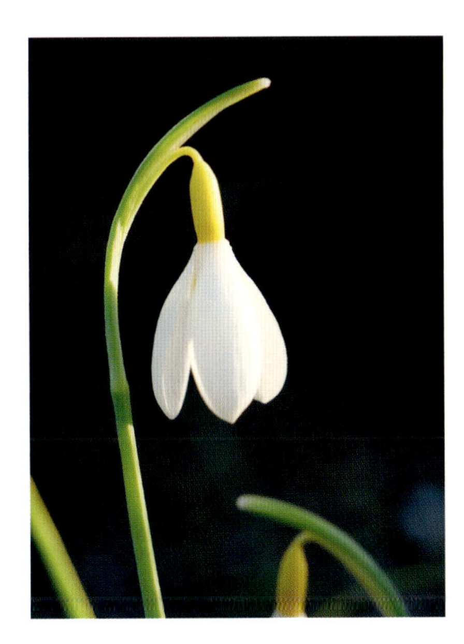

***Galanthus plicatus* 'Wendy's Gold'**

'Esther Merton' is a double-flowered snowdrop, its circular rows of inner segments marked with a green inverted V on both surfaces. The three outer petals are pure white. Height: 20 cm/8 in, flowers from January.

'Godfrey Owen' is a superb snowdrop that has six symmetrically arranged outer petals, and six inner ones that have green markings on each side of the small cut-out section at the tip of the inner petals, known as the 'apical notch'. Height: 20 cm/8 in, flowers from February.

'Grumpy' is pretty self-explanatory: two green dots for eyes and an inverted V for the mouth make a grumpy-looking face on the inner segments of the snowdrop. Height: 15–20 cm/6–8 in, flowers from January.

'Margaret Biddulph' is a striking snowdrop; its three inner segments are almost completely green on both sides, and the outer segments are washed with a paler green, sometimes forming stripes. Height: 15 cm/6 in, flowers from February.

GALANTHUS PLICATUS
– Pleated snowdrop

This snowdrop usually flowers slightly later than the previous two species and is useful for extending the season.

The flowers and leaves are larger than those of *G. nivalis*, but not as big as those of *G. elwesii*. The leaf shape also differentiates this species from the others, the pleated edges creating a central darker grey channel.

This species has produced many of the finest snowdrop cultivars available.

'Diggory' is very recognisable, with its pure white, wide, ballooning outer segments that have a puckered surface. The inner segments have pale green markings. Height: 15–20 cm/6–8in, flowers from February.

'Greenfinch' has green tips on the outer segments that are made up of small individual stripes, and a larger green marking on the inner segments. Height: 15–20 cm/6–8 in, flowers from early January.

'Percy Picton' holds its flowers on long arching pedicels that allow them to swing in the lightest of breezes. Height: 20 cm/8 in, flowers from late February.

'Trymlet' has green markings on both the outer and inner segments. The outer ones are slightly reflexed at the tips with crescent-shaped markings. Height: 15 cm/6 in, flowers from February.

'Wendy's Gold' is a popular yellow-marked cultivar. Its white nodding flowers are topped with a greenish-golden ovary and its inner segments are almost completely yellow. Height: 15 cm/6 in, flowers from February.

FOR THE BEST RESULTS SNOWDROPS should be planted 'in the green' soon after the flowers have faded. Large clumps can also be divided at this time to increase and spread plants around the garden. Planting dormant bulbs can be successful, as long as the bulbs have not become desiccated, which is often the case.

ASPECT: Best in partial shade
FLOWERING: January – March
HARDINESS: –10 to –15°C (14 to 5°F)

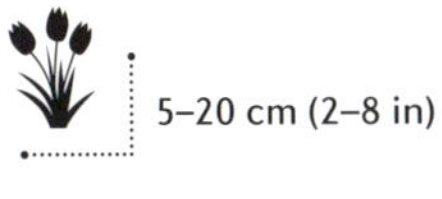

5–20 cm (2–8 in)

10 cm (4 in)

IRIS FOETIDISSIMA

Stinking iris

IRIDACEAE

An evergreen perennial, the stinking iris produces tall, broad, spear-like leaves from an underground rhizome, and is an ideal plant for growing in a shady spot under and close to trees. Its late spring flowers are a dull purple colour, but it is the showy fruits that the plant is most noted for. In late autumn, its large, ripe seed pods split open to reveal rows of bright orange–red fruits, which last through the winter months since birds do not appear to have a taste for them.

The plant gets its unfortunate common name 'stinking iris' from the odour emitted by its crushed leaves, which are said to smell 'beefy'. This also explains its other name, the roast beef plant.

ASPECT: Partial shade
FRUITS: November – February
HARDINESS: –10 to –15°C (14 to 5°F)

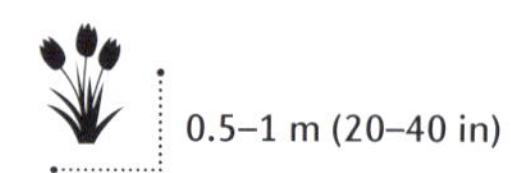

Iris foetidissima

IRIS RETICULATA

Early bulbous iris

IRIDACEAE

This pretty little fragrant iris grows from a small bulb surrounded by fibrous netting, from which it gets another of its common names, the netted iris. Its vibrant flower colours stand out when it's planted at the front of a border among snowdrops or winter aconites.

The fragrant flowers range in colour from pale blue to deep violet, and the blooms are set off by grass-like foliage. The outer petals have a lower lip, known as the 'fall', which is usually marked with white blotches or spots. It may also feature a yellow stripe or blotch.

'Fabiola' has blue flowers with a darker blue–purple fall, which features a white marking and a central short yellow stripe.

'Harmony' is royal blue with a white-marked fall with a central yellow stripe.

'Katharine Hodgkin' has distinctive creamy-white flowers covered in pale blue veining, with a yellow blotch on the fall. The flowers are slightly larger than some of the other early irises.

'Pauline' has dark purple flowers with white highlights, spotted with purple on the falls.

Iris reticulata does well when planted in a container, alone or with other plants, while growing it in a pot indoors is the best way to enjoy the fragrance.

ASPECT: Full sun
FLOWERING: January – March
HARDINESS: –10 to –15°C (14 to 5°F)

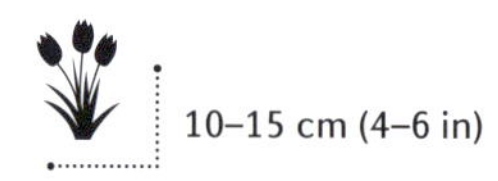

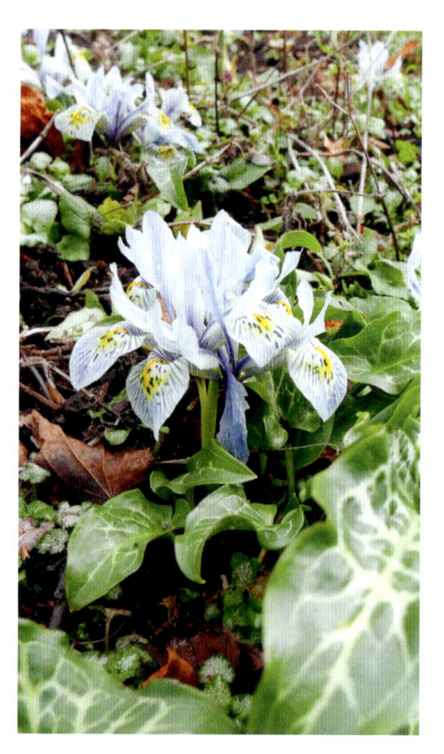

Iris reticulata

IRIS UNGUICULARIS

Algerian iris

IRIDACEAE

Unlike *Iris reticulata*, *Iris unguicularis* is evergreen. It grows from a rhizome that produces clumps of grass-like dark green leaves, from which large, lavender-blue, fragrant flowers with yellow-marked falls push up. The attractive flowers persist over a long period from late winter.

The foliage can sometimes become a bit untidy and can be cut back after flowering. Large clumps can be divided from mid- to late summer.

ASPECT: Full sun
FLOWERING: January – March
HARDINESS: –10 to –15°C (14 to 5°F)

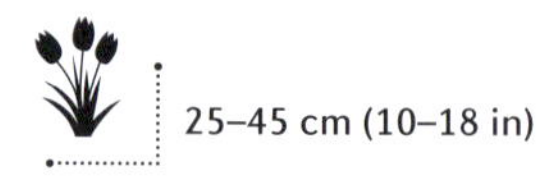

OPPOSITE
Iris unguicularis

LEUCOJUM VERNUM

Spring snowflake

AMARYLLIDACEAE

The flowers of the spring snowflake are usually produced before the leaves and are often mistaken for snowdrops, as they flower around the same time. Each stem of this decorative bulb carries a single or, occasionally, a pair of flowers that look like old-fashioned lamp shades, with six pointed petals with green or yellow tips.

Vernum means spring, but in most years this snowflake starts to flower a little earlier, just before the meteorological spring begins in mid-March.

Leucojums grow particularly well in moist but well-drained soils, where they will naturalise freely.

ASPECT: Full sun, partial shade
FLOWERING: February – March
HARDINESS: –10 to –15°C (14 to 5°F)

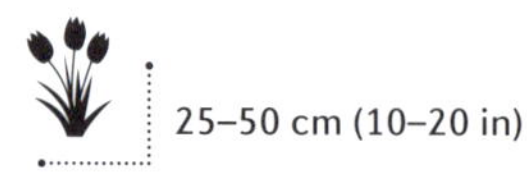

25–50 cm (10–20 in)

10 cm (4 in)

NARCISSUS BULBOCODIUM

Hoop petticoat daffodil

AMARYLLIDACEAE

This pretty little yellow wild daffodil has a wide-flaring yellow central trumpet surrounded by small star-shaped petals. It is commonly known as the hoop petticoat daffodil because of the flared shape of its corona, which resembles the flowing lines of an old-fashioned petticoat.

A bulbous perennial, its small size makes it ideal for growing in winter and spring containers, and it will also naturalise well in beds and grass, alongside snowdrops and crocuses. Like all bulbs that are naturalised in grass, you need to wait until the leaves have died back completely before mowing. This will allow the leaves to continue to feed the bulbs to produce flowers the following year.

There are various cultivars available, including Narcissus 'Golden Bells'.

ASPECT: Full sun
FLOWERING: February – March
HARDINESS: –10 to –15°C (14 to 5°F)

10–20 cm (4–8 in)

10 cm (4 in)

NARCISSUS CANTABRICUS

White hoop petticoat daffodil

AMARYLLIDACEAE

Another little wild daffodil, this bulb is easily recognised by the shape of the central trumpet (corolla), which also resembles a wide-flaring petticoat.

The species name *cantabricus* was given because it was originally, incorrectly, thought to have come from Cantabria in northern Spain, but, in fact, it is found in southern Spain and Morocco, where it grows naturally.

It is best planted in a sunny position so that the bulbs can ripen in the heat of the summer sun. However, it isn't grown that frequently as it's slow to naturalise and spread, but even in small numbers the pretty little white-flared trumpets are a joy to see in the garden in the depths of winter.

It is also a great bulb for growing in a container, either on its own or as part of a mixed winter display.

ASPECT: Full sun
FLOWERING: December – February
HARDINESS: –10 to –15°C (14 to 5°F)

10–15 cm (4–6 in)

10 cm (4 in)

NARCISSUS CYCLAMINEUS

Cyclamen-flowered daffodil

AMARYLLIDACEAE

An unusual-looking little yellow daffodil, this species flowers comprise a long, narrow tubular trumpet with a frilly margin that droops down towards the ground, while the petals (perianth segments) are totally reflexed and point up.

It grows well in grassy areas that hold

Spring snowflake is often confused with snowdrops, but the six-pointed, green or yellow tipped petals and lampshade-shaped flowers are very distinct

Leucojum vernum

Narcissus cyclamineus

Narcissus papyraceus

some moisture, where it will flower before the grass grows too long to hide it. It will naturalise in this situation but is quite slow to spread.

This bulb never used to flower until March, but now regularly blooms in February in the south of England, probably later in the north of the country.

ASPECT: Full sun, partial shade
FLOWERING: February – March
HARDINESS: –10 to –15°C (14 to 5°F)

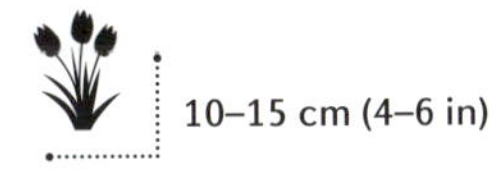

10–15 cm (4–6 in)

10 cm (4 in)

NARCISSUS PAPYRACEUS

Paper-white daffodil

AMARYLLIDACEAE

The pure white flowers of this elegant daffodil are produced in clusters of up to ten individual flowers, each with six petals and a small, central, cup-like corolla holding bright yellow stamens. In southern Spain it is often used as a cut flower at Christmas. The flowers are fragrant (but not to everyone's taste).

This amazing Mediterranean daffodil is hardy. I can remember when I first planted it in the garden, and the joy of seeing it bloom for the first time in December, a few weeks before Christmas. The following morning, I went out to photograph it only to find we had had snow overnight and there was no sign of the little daffodil at all.

A couple of days later the snow had gone and to my amazement there was the daffodil standing tall and in perfect condition. I have had the same group of plants in the garden now for more than ten years and they continue to perform year after year.

ASPECT: Full sun, to partial shade
FLOWERING: December – March
HARDINESS: –5 to –10°C (23 to 14°F)

20–40 cm (8–16 in)

5–10 cm (2–4 in)

NARCISSUS PSEUDONARCISSUS

Wild daffodil

AMARYLLIDACEAE

For me, there is something special about seeing the first of our native daffodils push through the soil in late winter. First, just the odd one appears and then, as the days start to lengthen slightly, they come on in a flourish, taking us from winter into spring. They are a flower that I associate more with spring, as that is their main flowering season, but they get started in late February or early March.

The glaucous strap-like leaves appear first, followed by the flowers, each bulb producing a single bloom. The central deep yellow trumpet flares out towards its frilly opening and is surrounded by six pale yellow twisting petals (perianth segments).

These daffodils make great cut flowers to bring into the house, grow well in pots and are ideal for naturalising in long grass.

ASPECT: Full sun, partial shade
FLOWERING: February – April
HARDINESS: –10 to –15°C (14 to 5°F)

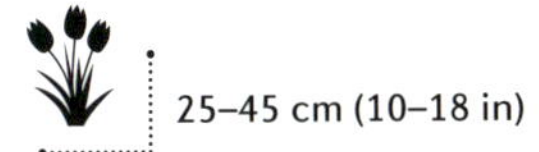

Narcissus cultivars

'February Gold', as the name suggests, begins to flower in February, with deep yellow flowers that feature distinctive swept-back petals and a long trumpet (corolla). Ideal for naturalising. Flowers from February to March.

'February Silver' has white flowers with a pale lemon trumpet (corolla). Flowers from February to March.

'Rijnveld's Early Sensation' is probably the earliest of the cultivars to flower, blooming in December in a mild winter. The bright yellow flowers have a short, wide, tapering trumpet (corolla). Flowers from January to March.

ASPECT: Full sun to full shade, best in partial shade
FLOWERING: December – March
HARDINESS: –10 to –15°C (14 to 5°F)

PUSCHKINIA SCILLOIDES VAR. *LIBANOTICA*

Russian snowdrop

ASPARAGACEAE

Also known as the Russian snowdrop, or Lebanon striped squill, this little bulbous perennial produces a single flower spike from each bulb that holds up to ten star-shaped flowers on individual short stems. The pale blue flowers have six petals, each with a central darker stripe.

The Russian snowdrop will spread slowly if left undisturbed, and is a great little plant for naturalising under deciduous trees and shrubs.

Puschkinia libanotica

There is also a pure white form, ***Puschkinia scilloides* var. *libanotica* 'Alba'**.

ASPECT: Full sun, partial shade
FLOWERING: February – April
HARDINESS: –10 to –15°C (14 to 5°F)

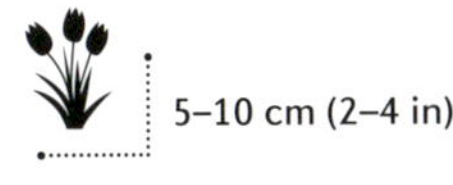

This golden-yellow daffodil is one of the earliest, and has a long flowering period

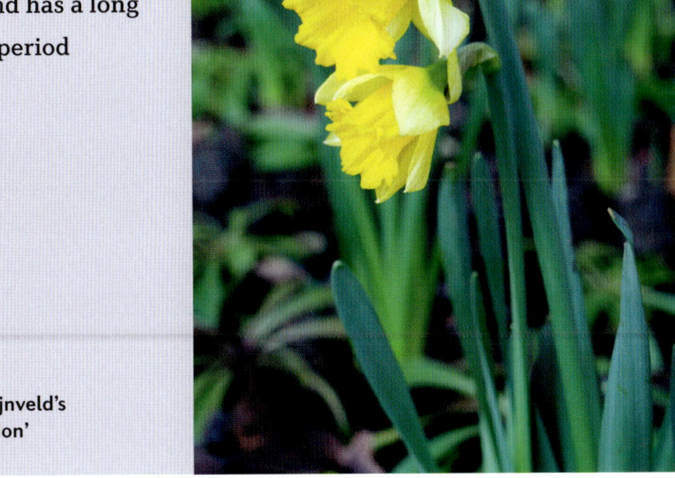

***Narcissus* 'Rijnveld's Early Sensation'**

OPPOSITE
Narcissus pseudonarcissus

Climbers

CLEMATIS ARMANDII

Evergreen clematis

RANUNCULACEAE

A vigorous evergreen climbing plant, this clematis has glossy, leathery, lance-shaped, dark green leaves, and will cover a fence or wall in a very short space of time, given the support of a trellis or wires for its twining tendrils to attach to.

The flowers appear in late winter and last into early spring. The six-petalled, fragrant, pure white blooms are so prolific that at their peak they almost obscure the foliage while filling the air with their vanilla-like scent.

It is a good idea to prune the evergreen clematis after flowering once it has filled its allotted space, which will not only improve flowering the following year, but also clear out a lot of the dead leaves that get trapped among its stems.

Although the plant itself is very hardy, when in full flower a heavy frost will damage the petals, turning them brown.

ASPECT: Full sun, partial shade

FLOWERING: February – April

HARDINESS: –5 to –10°C (23 to 14°F)

3–8 m (10–25 ft)

2–3 m (6–10 ft)

Clematis cirrhosa

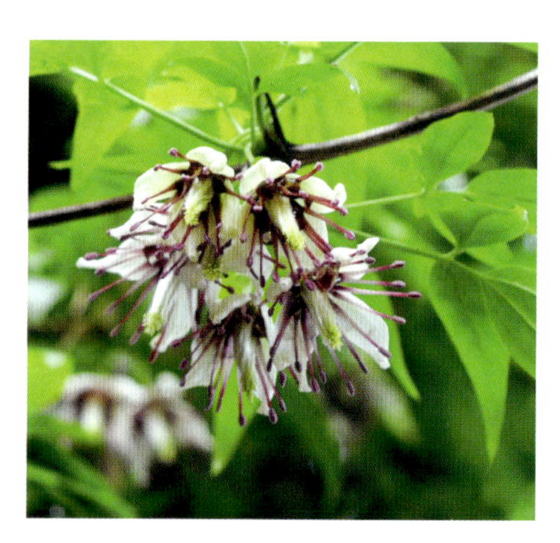

Clematis napaulensis

CLEMATIS CIRRHOSA

Winter clematis

RANUNCULACEAE

Probably my favourite winter-flowering clematis, I first saw it growing naturally in the wild in southern Spain, where both the creamy-white and the speckled forms can are found scrambling through trees and bushes. There are also some selected forms that have been given cultivar names.

Masses of fragrant, four-petalled, bell-shaped flowers hang on long trailing stems, alongside the glossy green leaves. These are followed by attractive silky seedheads, which sparkle in the winter and early spring sunshine.

A general tidy by removing dead or damaged growth after flowering is all the care it needs. This is also a very drought-tolerant plant, being native to the Mediterranean, so requires little watering throughout the summer once established. Overwatering at this time will actually be detrimental.

Two of the most popular cultivars are **'Freckles'**, which is heavily speckled with reddish markings inside each petal, and **'Jingle Bells'**, which is closer to the wild species and features creamy-white petals.

ASPECT: Full sun, partial shade

FLOWERING: November – March

HARDINESS: –5 to –10°C (23 to 14°F).

2–3 m (6–10 ft)

11.5 m (3–5 ft)

Best planted in a sheltered spot for protection, and where the pretty bell-shaped flowers will brighten a shady area

Clematis 'Winter Beauty'

CLEMATIS NAPAULENSIS

Nepal clematis

RANUNCULACEAE

Native to Nepal and parts of China, this is a relatively rare clematis in cultivation. The flowers are very showy, and it is unusual in that it drops its leaves in late summer when it undergoes a short period of dormancy.

It starts to come back into leaf in late autumn, producing bright green leaves that are divided into three or five smaller leaflets. Its nodding bell-shaped flowers are borne in small clusters, each held on a long stem, and comprise four creamy-white or pale yellow–green petals with tips that become slightly reflexed with age. The blooms also feature long pinkish stamens and prominent reddish-purple anthers, and large fluffy seedheads appear after flowering.

It needs a bit of protection and is best planted in a sheltered position – it may not be hardy in cold regions that regularly experience prolonged periods below freezing.

Any pruning should be carried out after flowering in early spring.

ASPECT: Full sun, partial shade
FLOWERING: December – March
HARDINESS: 1 to –5°C (34 to 23°F)

2–4 m (6–12 ft)

1–1.5 m (3–5 ft)

CLEMATIS UROPHYLLA 'WINTER BEAUTY'

Clematis 'Winter Beauty'

RANUNCULACEAE

A vigorous evergreen climber, once established, the dark green foliage looks good all year round, while clusters of nodding flowers appear in the depths of winter. The fragrant blooms are the palest green, becoming pure white and bell-shaped with age, each borne on a pinkish stem. The thick waxy petals also develop flared tips when fully open, when they reveal a boss of creamy anthers. These attractive flowers stand out well against the dark leaves.

Plant this clematis in a sheltered spot, out of the worst cold winter winds and frost, where it will quickly cover a wall or fence. It is fairly low maintenance, but any necessary pruning should be carried out immediately after flowering.

ASPECT: Full sun, partial shade
FLOWERING: December – March
HARDINESS: 1 to –5°C (34 to 23°F)

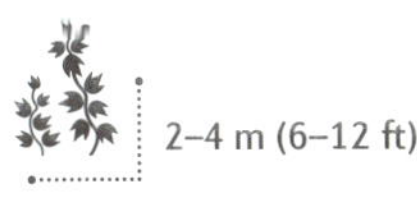

2–4 m (6–12 ft)

1–2.5 m (3–8 ft)

HEDERA
The ivies

This invaluable group of plants is suitable for growing year-round but they have a special place in a winter garden. Whether you grow them on a wall, fence, or trellis, as scrambling ground cover, or use them in a container such as a hanging basket or window box, ivies are suited to a range of situations. Not very fussy about soil conditions (though they do best in alkaline soils), as long as their roots are not too wet, they will grow well in full sun or full shade. Those with yellow variegation will produce the best colour in a sunny position, while the silver-variegated types will be happy in a shady or bright spot.

Ivies are also great plants for wildlife, offering food and a home for many creatures throughout the year. A haven for nesting birds in spring and summer, as well as roosting bats, they also provide a late source of nectar and pollen in early autumn for many insects, particularly bees and butterflies. Their black berries are a good source of winter food, too, eaten by birds during the coldest months, and their network of stems and leaves also provide a warm and dry home for over-wintering insects.

Only a few are mentioned here, which have been chosen for their winter hardiness and their reliability, but there are many different types to choose from, featuring a range of leaf colours, shapes, and sizes.

Pruning can be carried out whenever needed to keep plants compact but it is best to cut these climbers in early spring, before nesting begins, if you are growing them with wildlife in mind.

The unique patterns of each each leaf are a mix of creamy, whites, yellows, and silvery-greys

***Hedera algeriensis* 'Gloire de Marengo'**

It is also worth remembering that the self-clinging aerial roots of all ivies may cause some damage to walls with unsound mortar.

Gloves should be worn when pruning as the sap can be an irritant to some.

HEDERA ALGERIENSIS 'GLOIRE DE MARENGO' - Algerian ivy - ARALIACEAE
This cultivar of Algerian ivy is easy to recognise, with its large, variegated leaves growing along reddish-purple stems. Each dark green leaf is decorated with a unique pattern made up of a range of colours that include cream, grey, silver and yellow, while the new leaves have the palest primrose-yellow margins.

The leaf shape is variable too: while most are lobed, you may find some that are unlobed and heart-shaped all on the same plant.

'Margino Maculata' has speckled variegation all over the leaves in a mixture of cream, green, and grey. It is slightly less hardy than 'Gloire de Marengo'.

Like all ivies, it is self-clinging and evergreen but this one needs a sheltered spot out of cold winter winds. It can also be grown as a house plant, or in a conservatory in a large container.

Pruning can be carried out at any time of year.

ASPECT: Full sun, partial shade
FOLIAGE INTEREST: Year round
HARDINESS: −10 to −15°C (14 to 5°F)

2.5–4 m (8–12 ft)

HEDERA COLCHICA

Persian ivy

ARALIACEAE

The Persian ivy is a vigorous evergreen self-clinging climber. Its large leaves are heart-shaped and glossy green, and emit a lemony fragrance when crushed. Although the plain green-leaved species has its place in a garden, particularly for screening, it is the variegated forms that will bring a touch of light and colour to a dark wall, fence, or corner, and are probably the most useful during the grey days of winter. This ivy's adventitious roots do not grip on to walls as well as those of some of the smaller-leaved species, so it is not ideal for growing on taller structures, or on the walls of houses, as strong winds can bring it down. However, the Persian ivy is more tolerant of acidic soils than the common ivy, *Hedera helix*.

Two of the best cultivars are:

'Dentata Variegata', which has creamy-yellow margins with mottled grey–green leaves.

'Sulphur Heart' has dark green leaves with a central splash of gold or paler green.

ASPECT: Full sun, partial shade
FOLIAGE INTEREST: Year round
HARDINESS: –10 to –15°C (14 to 5°F)

6–8 m (20–25 ft)

4–6 m (12–20 ft)

HEDERA HELIX

Common ivy

ARALIACEAE

The foliage of common ivy is very variable, and while it is generally dark green with three to five lobes, the leaves are often enhanced with a lighter-coloured midrib and veins. They may also have bronze tints in autumn.

Hedera helix

Common ivy also produces two types of leaf: juvenile and mature. The younger foliage has adventitious aerial roots that help the plant to cling to most surfaces, while the mature leaves are heart-shaped or oval and do not have the lobes of the juvenile foliage, or the self-clinging aerial roots. But it is the mature ivy stems that bear the yellowish-green flowers, which go on to produce rounded clusters of black berry-like fruits.

This ivy is also of great value to wildlife. Nectar and pollen produced in late autumn provide the last and essential food sources for many overwintering insects, while the berries have a high fat content that helps to supplement the diet of many birds at this time of year.

Hedera helix is not very tolerant of acidic soils and will grow much better in alkaline conditions. There are more cultivars of Hedera helix than of any other ivy, and the list below is just a tiny selection to show the variability.

'Buttercup' is best grown in full sun where the buttery-yellow leaves will retain their colour. When grown in shade they become paler and greener. Height: 2–3 m/6–10 ft.

'Duckfoot' is a small, compact ivy, with leaves, as its name suggests, similar in shape to a small duck's foot: three-lobed and wedge-shaped at the base. Its size makes it ideal for growing in a container, and as a house plant. Height: 50 cm/20 in.

'Goldchild' has typical ivy-shaped leaves, with a green two-toned section in the centre and pale golden-yellow variegation at the edges. Height: 2–3 m/6–10 ft.

'Ice Cream' is a non-climbing form and produces ovate, pointed green leaves with white margins that have a touch of pink during the coldest months of the year. This clone produces flowers in late autumn followed by fruits in early winter. Height: 0.5–1 m/20–39 in.

'Midas Touch'

'Buttercup'

'Tripod'

'Goldchild'

'Maple Leaf'

'Duckfoot'

'Maple Leaf' has green leaves with five long, narrow, deeply cut lobes; the central lobe is the longest. This cultivar is particularly well suited to shade. Height: 2 – 4 m/6–12 ft.

'Midas Touch' has boldly coloured, heart-shaped leaves that turn under slightly along the margins. The combination of striking green and yellow variegation looks like the colours have been applied from an artist's palette. Height: 1–2 m/3–6 ft.

'Tripod' is a vigorous ivy with small, three-lobed, green leaves that have a paler green midrib and veins. Its central lobe is much longer than the lateral lobes, which almost point out at right angles to form a T shape. Height: 2–4 m/6–12 ft.

Pruning advice and precautions are the same as for the Persian ivy, *Hedera colchica*.

ASPECT: Full sun, partial shade
FOLIAGE INTEREST: Year round
HARDINESS: –10 to –15°C (14 to 5°F)

6–10 m (20–30 ft)

2–5 m (6–15 ft)

HEDERA HIBERNICA
Irish ivy

ARALIACEAE

For a long time, this ivy was thought to be a subspecies of *Hedera helix*, but analysis of their DNA has been proved that they are closely related, but separate species. Like *Hedera helix* Irish ivy is very variable, but it is more tolerant of acidic soils.

'Rona' is one of the showiest cultivars of Irish ivy, its foliage a patchwork of speckled creams, yellows, and greens. The youngest leaves display the best colour. As they age, they become greener, but it is the variation in colours produced by foliage of different ages that makes this ivy so attractive.

Pruning advice and precautions are the same as for Persian ivy, *Hedera colchica*.

ASPECT: Full sun, partial shade
FOLIAGE INTEREST: Year round
HARDINESS: –10 to –15°C (14 to 5°F)

OPPOSITE
***Hedera helix* 'Ice Cream'**

***Hedera hibernica* 'Rona'**

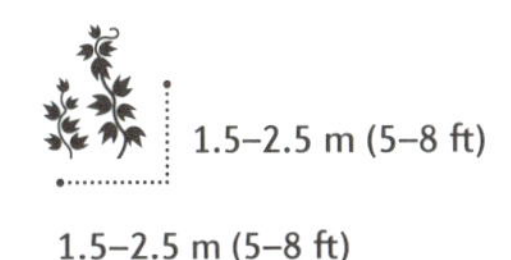

1.5–2.5 m (5–8 ft)

1.5–2.5 m (5–8 ft)

TRACHELOSPERMUM ASIATICUM 'SUMMER SUNSET'

Chinese jasmine

APOCYNACEAE

An evergreen, twining woody climber, the Chinese jasmine will need some supporting wires to help it climb a wall or trellis. Its glossy leaves are very variable, with golden-yellow splashes of colour in the middle of each leaf and dark green edges.

As well as making a colourful screen, this climber has the added bonuses of fragrant, creamy, star-shaped flowers in late summer, and bright red new growth in spring.

It is best grown in a sheltered spot as it is likely to be damaged by heavy frosts in a severe winter.

Although the common name might suggest that this is a jasmine, it is actually from a different family, although *Jasminum officinale* (common jasmine) and *Trachelospermum* share many characteristics.

Pruning is generally not needed but it can be trimmed back quite hard after flowering if it outgrows its space. Take care, though, as the sap contains latex, which is an irritant to some.

ASPECT: Full sun, partial shade
FOLIAGE INTEREST: Year round
HARDINESS: –5 to –10°C (23 to 14°F)

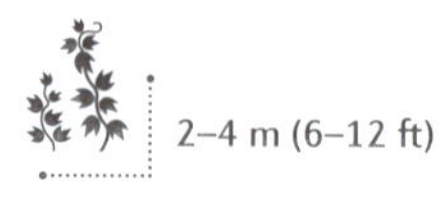

Trachelospermum asiaticum **'Summer Sunset'**

Trachelospermum jasminoides 'Variegatum'

Tropaeolum tricolor

TRACHELOSPERMUM JASMINOIDES 'VARIEGATUM'

Variegated star jasmine

APOCYNACEAE

Also known as the variegated Confederate jasmine, for most of the year it has glossy green leaves with creamy-white blotches and margins, but in the winter the leaves become tinged with red and pink.

It is best grown on a sunny, sheltered wall or fence where it will need a trellis or wires for its twining stems to climb through. This star jasmine can be a bit slow to get going but will grow quite quickly once established. Fragrant white tubular flowers are produced from mid- to late summer.

Any pruning that is needed can be carried out in early spring by reducing the length of the twining stems. Wear gloves to avoid skin contact with the latex in the sap.

ASPECT: Full sun, partial shade
FOLIAGE INTEREST: Year-round
HARDINESS: –5 to –10°C (23 to 14 °F)

1– 2.5 m (3–8 ft)

1–2.5 m (3–8 ft)

TROPAEOLUM TRICOLOR

Three-coloured nasturtium

TROPAEOLACEAE

This is an amazing, vigorous climber and one to try only if you have a very sheltered spot in the garden, or a conservatory, as it will not survive a frost.

I have often seen it growing in evergreen hedges such as yew (where the main stems of this perennial are protected from freezing conditions), the dark leaves helping to show off the unusual flowers, which are red, with a deep purple mouth and primrose-yellow centre. The blooms appear on and off in milder periods throughout winter.

It definitely creates a talking point when grown in a container up a trellis or wires in a frost-free conservatory.

This plant grows from a summer-dormant tuber, when it dies back and the stems need to be cut back to ground level. New growth will appear each year in September/October.

ASPECT: Full sun, partial shade
FLOWERING: February – April
HARDINESS: 1 to 5°C (33 to 41°F)

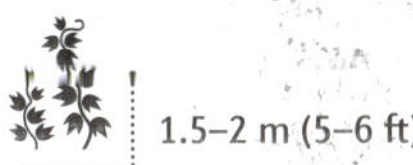

1.5–2 m (5–6 ft)

50 cm (20 in)

Ferns

Ferns that are green all year are officially classed as 'wintergreen', as opposed to 'evergreen'. This means that they keep their fronds through the winter months, but are dormant in summer, during which time the fronds naturally break away from the rhizome. However, they are usually called evergreen so I will refer to them as such in their descriptions.

They are probably not the first plants that spring to mind when you are planning a winter garden, but they are very versatile, with many different types and cultivars that will suit most gardens. The evergreens, in particular, make ideal plants for growing in shady areas.

Commonly called woodland ferns, their feathery, divided leaves are known as fronds, and, in some forms, they are split into pairs of oppositely arranged leaflets, which are called pinna. Others, such as the hart's tongue fern, have individual undivided or entire fronds. With a wide range of sizes, shapes, and textures to choose from, ferns will add interest when grown either in groups or in combination with other herbaceous plants such as hellebores and woodland bulbs, including snowdrops and winter aconites.

Many of the smaller varieties make good plants for growing in mixed winter containers and require very little maintenance.

ASPLENIUM SCOLOPENDRIUM

Hart's tongue fern

ASPLENIACEAE

The evergreen hart's tongue fern gets its name from the old English word for a male red deer or stag, and its tongue-like fronds.

The plant forms a rosette of strap-shaped fronds, or blades, each with slightly undulating margins and rows of sunken lines each side of the central rib. These are caused by the sporangia, the cases containing the spores (fern seeds), from which new plants are produced.

Naturalistic planting schemes suit these ferns best, in shady areas, tucked against walls, around rocks and with spring bulbs. They prefer alkaline soils but will tolerate most soil types.

Blechnum cordatum

Asplenium scolopendrium Crispum Group includes cultivars that have fronds with a distinctive wavy edge, sometimes along the whole length, or just a crested tip.

ASPECT: Partial shade, full shade

FOLIAGE INTEREST: Year round

HARDINESS: –10 to –15°C (14 to 5°F)

30–60 cm (12–24 in)

10–50 cm (4–20 in)

BLECHNUM CORDATUM

Chilean hard fern

ASPLENIACEAE

Until recently this tall perennial evergreen fern was known as *Blechnum chilense* and will probably be found for sale under that name for some time. The large, arching, leathery, glossy green fronds are produced in abundance from a central point, coming to life when lit up by winter sunshine.

Like all *Blechnum*, this attractive species has two types of frond, which are either sterile or fertile, known in botany as dimorphism. The majority are sterile, while the fertile fronds are taller, darker and narrower and carry the spores (see above).

Asplenium scolopendrium

This fern makes a good tall ground-cover plant that spreads by long creeping rhizomes.

The fronds can be cut down in spring to reveal salmon-coloured new growth.

ASPECT: Full sun, partial shade
FOLIAGE INTEREST: Year round
HARDINESS: –10 to –15°C (14 to 5°F)

0.9–1.5 m (3–5 ft)

50–75 cm (20–30 in)

BLECHNUM SPICANT

Hard fern

ASPLENIACEAE

This is a great little fern that can often be seen growing wild in the countryside. It is a British native but has a very wide range and can be found growing wild in the semi-shade of woodland edges and banks in much of the temperate world.

The hard fern's preference for woodland conditions makes it ideal for growing with winter- and spring-flowering bulbs in a naturalistic planting, especially in combination with snowdrops, but it also grows well in more open aspects.

This evergreen has upright, glossy, feather-like green fronds that are produced in a whorl and, like all *Blechnum* species, it has two different types of frond: sterile and fertile (see above). It prefers acid to neutral soil and can also be grown in ericaceous compost as part of a mixed planting in a large container.

ASPECT: Partial shade, full shade
FOLIAGE INTEREST: Year round
HARDINESS: –10 to –15°C (14 to 5°F)

10–60 cm (4–24 in)

10–50 cm (4–20 in)

Dryopteris affinis

DRYOPTERIS AFFINIS

Scaly male fern

DRYOPTERIDACEAE

A semi-evergreen fern that produces largish clumps of long and wide-spreading golden-green fronds when young, covered in golden-russet scales along the underside of the spine (botanically known as the stipe) of each one. The fronds then darken as they mature, and the plant produces a beautiful shuttlecock-like shape. Tolerant of both alkaline or acid soils, it will grow happily in sun or shade, and once established is also very drought tolerant. There are also a number of cultivars with wavy and crested margins. For example, *Dyropteris* **'Cristata'**, or **'Crispa'**.

ASPECT: Full sun, partial shade
FOLIAGE INTEREST: Year round
HARDINESS: –10 to –15°C (14 to 5°F)

1–1.5 m (3–5 ft)

0.5–1 m (20 in–3 ft)

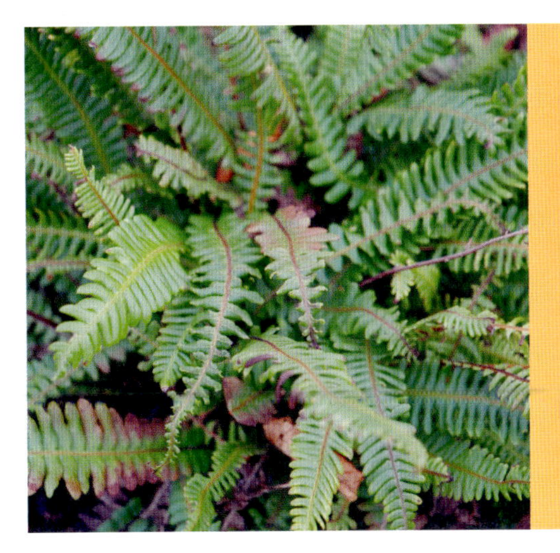

Ideal for growing in a semi-shaded area combined with early-flowering bulbs

Blechnum spicant

Dryopteris erythrosora **'Brilliance'**

DRYOPTERIS ERYTHROSORA 'BRILLIANCE'

Copper shield fern

DRYOPTERIDACEAE

The copper shield fern, or autumn fern, is semi-evergreen, so the foliage will not generally survive all winter but in late autumn the red spores on the back of each frond offer good late-season colour. The fronds last until the severe frosts arrive in mid-winter, sparkling wonderfully when lightly dusted with ice crystals.

Dryopteris erythrosora **'Brilliance'** is a particularly good cultivar, with bright gold and coppery-toned young fronds, which retain some of these tones even when mature, enhanced by the orangey-red spores on the undersides of the fronds in late autumn.

Once the foliage becomes untidy, it can be cut down to ground level and will produce striking new orange-red fronds in early spring alongside the last of the winter bulbs.

ASPECT: Partial shade, full shade
FOLIAGE INTEREST: Year round
HARDINESS: –10 to –15°C (14 to 5°F)

DRYOPTERIS WALLICHIANA

Alpine wood fern

DRYOPTERIDACEAE

This tall, impressive fern is semi-evergreen and forms clumps of wide, lance-shaped fronds that grow in the shape of a giant shuttlecock.

It is best grown in a shady or woodland-type setting where it will thrive, adding foliage interest during the winter months and a backdrop to early-flowering snowdrops such as *Galanthus nivalis*. The fronds may be damaged by heavy snowfall and severe frost, but when planted among hardy shrubs and deciduous trees that offer protection from the coldest winds, it will sparkle and glisten throughout the season. Remove dead and damaged fronds as they appear. Large clumps can be cut back before the new fronds emerge, and divided, if necessary, in spring.

ASPECT: Full sun, partial shade
FOLIAGE INTEREST: Year round
HARDINESS: –10 to –15°C (14 to 5°F)

0.5–1 m (20–39 in)

0.5–1 m (20–39 in)

Polystichum setiferum

Polypodium vulgare

POLYPODIUM VULGARE

Common polypody

POLYPODIACEAE

A very hardy evergreen fern that can tolerate many different conditions from dry, bright, sunny spots to cool, dappled shade. It can also be found growing naturally both in the ground (terrestrial), and colonising tree branches as an epiphyte.

This polypody copes well with winter conditions, too, both in containers and as part of a mixed planting under and around deciduous trees and shrubs, together with hellebores and winter bulbs, where it will slowly spread via its creeping rhizomes.

Its evergreen, ladder-like, leathery fronds have distinctive raised pimply sporangia (see *Asplenium scolopendrium*) that form parallel lines on the undersides of each of the leaflets. In very cold weather the fronds curl up to reduce their leaf surface area and look particularly handsome when their edges are highlighted with frost.

To keep these ferns looking good, simply remove any damaged or dead fronds in spring.

ASPECT: Full sun, partial shade
FOLIAGE INTEREST: Year round
HARDINESS: –10 to –15°C (14 to 5°F)

10–30 cm (4–12 in)

10–50 cm (4–20 in)

POLYSTICHUM SETIFERUM

Soft shield fern

DRYOPTERIDACEAE

This fern is native to Britain and looks good when grown with other ferns that have a different habit such as the hart's tongue fern, *Asplenium scolopendrium*.

It's very hardy but can become a bit tatty if exposed to strong winds. When given the protection of hardy shrubs and trees, or grown on a sheltered bank, it makes a good companion for late winter bulbs and perennials such as snowdrops and primroses.

Remove any fronds that have become shabby when necessary.

ASPECT: Full sun, partial shade
FOLIAGE INTEREST: Year round
HARDINESS: –10 to –15°C (14 to 5°F)

0.5–1 m (20–39 in)

0.5–1 m (20–39 in)

Grasses

Ornamental grasses are an invaluable addition to the winter garden, offering many different colours and textures, while providing movement as they sway gracefully in the gentlest of breezes. They come in a range of colours, including warm bronze, coppery red, golden-yellow and green, or you can opt for one of the variegated forms.

Most grasses have year-round appeal, which is a bonus, offering interest throughout the seasons, not just winter. Those with large flowerheads such as fountain grass (*Pennisetum*) look great when backlit by the low winter sun, while a touch of frost transforms them into sparkling jewels.

In the list below I have also included a few of the grass-like sedges. Many of these are low-growing evergreens that come in a variety of colours and need very little, if any, maintenance.

Grasses and grass-like plants have so many uses at this time of year. You can include them as a backdrop to other plants in beds or borders or use them as standalone specimens. The taller types also make ornamental divisions or screens. However you use them, grasses are sure to make an impact.

A note of caution: always wear gloves when cutting back or removing dead foliage from grasses and sedges, as many have very sharp edges that may cut you.

ACORUS GRAMINEUS 'OGON'

Slender sweet flag

ACORACEAE

This pretty little grass-like perennial is evergreen in all but the coldest of winters, so may not be suitable for areas of the country that experience prolonged and regular periods of hard frost, if you want it to perform in a seasonal display.

***Acorus gramineus* 'Ogon'**

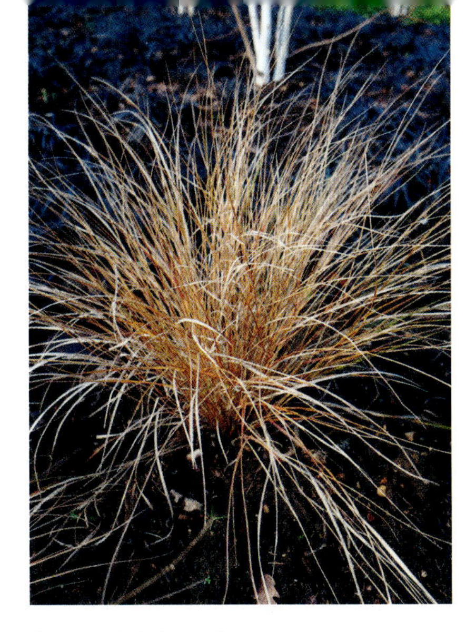

Anemanthele lessoniana

However, at worst, the frost will damage the foliage, since the plant itself is fully hardy, so it should grow back perfectly well in the spring.

The beautiful golden-yellow variegated leaves of 'Ogon' make great ground cover and the plant will do particularly well when grown in partial or semi-shade around shrubs, but it looks equally good in a drift on its own or combined with other perennials such as hellebores.

Remove dead and damaged leaves, when necessary, by running a rake through the foliage.

ASPECT: Full sun, partial shade
FOLIAGE INTEREST: Year round
HARDINESS: –10 to –15°C (14 to 5°F)

 **10–50 cm (4–20 in)**

10–50 cm (4–20 in)

ANEMANTHELE LESSONIANA
Pheasant's tail grass

POACEAE

Formerly known as *Stipa arundinacea*, this evergreen grass performs year round, producing clumps of loose, arching foliage with pink flowerheads in late summer, then, as the seasons move into autumn, the leaves take on tints of orange, copper, and gold. The colours become even more intense during the winter when this beautiful grass really stands out. It isn't too fussy about the soil conditions and will thrive in a variety of garden situations. Try growing it in drifts through other plants or in a large container. Divide large clumps in late spring and early summer.

This grass produces seed freely so will spread around; remove any self-sown seedlings that pop up in unwanted areas or pot them up and grow on to replace older plants.

ASPECT: Full sun, partial shade
FOLIAGE INTEREST: Year round
HARDINESS: –5 to –10°C (23 to 14°F)

 **50–90 cm (20–36 in)**

0.5–1 m (20 in–3 ft)

CALAMAGROSTIS BRACHYTRICHA
Korean feather reed grass

POACEAE

Tall and elegant, this slow-spreading ornamental grass flowers late in the season, producing large, fluffy, pinkish flowerheads in autumn that persist into the winter, when they become golden-coloured. They look particularly beautiful when backlit by the low winter sun. The grey–green summer foliage also takes on a soft light tan colour at this time of year, which complements the seedheads.

Calamagrostis brachytricha isn't an evergreen grass so it needs to be cut back in late winter or early spring before the new growth appears.

ASPECT: Full sun, partial shade
INTEREST: Year round
HARDINESS: –10 to –15°C (14 to 5°F)

 **1–1.5 m (3–5 ft)**

50 cm–1m (20 in–3 ft)

CALAMAGROSTIS X ACUTIFLORA
'Karl Foerster' Feather reed grass

POACEAE

This tall, architectural, deciduous grass is in many ways similar to the Korean feather reed grass, but its stiffer stems are less prone to flopping in exposed areas of a garden and it makes a good screen when planted en masse as a backdrop to brighter-coloured plants such as dogwoods (*Cornus*).

During the autumn the feathery pinkish-purple flowerheads of late summer turn to golden-tan seedheads and, along with the straw-coloured stems and leaves, they remain attractive throughout the winter.

This is another grass that needs to be cut back in late winter or early spring before the new growth appears.

ASPECT: Full sun, partial shade
INTEREST: Year round
HARDINESS: –10 to –15°C (14 to 5°F)

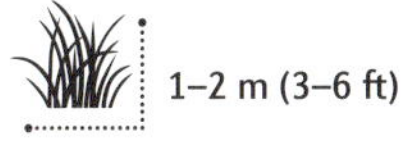

50 cm– 1 m (20 in–3 ft)

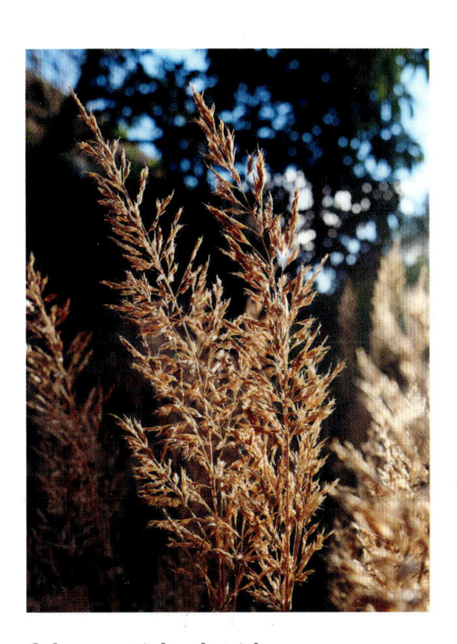

Calamagrostis brachytricha

CAREX COMANS 'BRONZE-LEAVED'

Bronze New Zealand hair sedge

CYPERACEAE

A very decorative bronze-coloured evergreen sedge, this species produces dense tussocks of narrow, arching, grass-like leaves. It is ideal for planting at the edge of a bed or border and around the base of deciduous shrubs. It also tolerates some shade, but prefers to grow in full sun, where it will also develop a more vibrant colour. In summer, it also produces inconspicuous brown flower spikes hidden among the foliage.

Because of its small size it is an ideal grass for growing in a container along with winter-flowering plants, particularly spring bulbs.

ASPECT: Full sun, partial shade
FOLIAGE INTEREST: Year round
HARDINESS: –5 to –10°C (23 to 14°F)

10–30 cm (4–12 in)

10–40 cm (4–16 in)

Carex comans

CAREX MORROWII 'VARIEGATA'

Japanese sedge

CYPERACEAE

The vivid colours of this variegated sedge will brighten up a shady border edge and, like most sedges, which grow naturally in damp areas, it thrives in moist but well-drained soil and tolerates wetter conditions better than most grasses and other grass-like plants.

It forms small, tufted mounds of strap-like bright green leaves with creamy-white bands both on the leaf margins and on the central area.

Grown in a group or with darker contrasting foliage plants such as *Bergenia purpurascens* or a dark-leaved *Ajuga*, it makes a standout winter plant. It is also a striking container plant, adding colour and texture to a cold season display.

This sedge is slow to spread, but, if needed, divide it in late spring.

ASPECT: Full sun, partial shade
FOLIAGE INTEREST: Year round
HARDINESS: –10 to –15°C (14 to 5°F)

30–45 cm (12–18 in)

30–60 cm (12–24 in)

CAREX OSHIMENSIS 'EVERGOLD'

Japanese sedge

CYPERACEAE

A small evergreen clump-forming plant, this sedge produces long, arching, dark green leaves, each decorated with a thick, central, creamy-yellow stripe, which from a distance gives the plant an eye-catching golden appearance.

The small size again makes this little sedge ideal for growing in containers with other brightly coloured plants, where it can help to illuminate a shady spot on a patio or courtyard.

Carex oshimensis 'Evergold'

It just requires a tidy in summer, when any dead leaves should be removed.

ASPECT: Full sun, partial shade
FOLIAGE INTEREST: Year round
HARDINESS: –10 to –15°C (14 to 5°F)

25–30 cm (10–12 in)

30–40 cm (12–16 in)

CAREX TESTACEA

Orange New Zealand sedge

CYPERACEAE

Similar to *Carex comans* in colour and form, this sedge is a taller plant but still small enough to grow close to the front of a border. Its foliage turns from olive-green in summer to a coppery-brown in the autumn, after which the leaves take on orange tints during the winter, making it a very attractive ground-cover plant all year round.

The cultivar **'Prairie Fire'** also forms low arching mounds of leaves, but their olive-green colour is infused and tipped with orange highlights.

Like *C. comans*, this sedge is perfect for growing in containers with other winter interest plants.

It can be divided in early summer.

ASPECT: Full sun, partial shade
FOLIAGE INTEREST: Year round
HARDINESS: –10 to –15°C (14 to 5°F)

30–45 cm (12–18 in)

30–60 cm (12–24 in)

DESCHAMPSIA CESPITOSA
Tufted hair grass

POACEAE

This elegant grass looks very different in summer and winter. During the summer months, masses of cloud-like, feathery flowers completely mask the stems, but as the season turns to autumn, the billowing blooms evolve into bronze-brown seed heads held on similarly coloured stems that persist well into winter. These look particularly beautiful when swaying in the breeze.

The light, airy appearance of this grass lends it to growing among shrubs or other perennials such as *Phlomis russeliana* grown for winter seed heads.

Deschamsia needs to be cut back in late winter or early spring before the new growth appears.

ASPECT: Full sun, partial shade
INTEREST: Year round
HARDINESS: –10 to –15°C (14 to 5°F)

0.5–1 m (20–39 in)

50 cm–1 m (20 in–3 ft)

FESTUCA GLAUCA 'ELIJAH BLUE'
Blue fescue

POACEAE

Probably the best and most reliable of the blue fescue grasses, 'Elijah Blue' is widely grown for its silver–blue foliage.

This dwarf ornamental grass needs to be grown in full sun to maintain the best blue colour and will appear greener when planted in shade. It forms compact rounded mounds of needle-like leaves, giving it a spiky appearance that contrasts well with many other plants.

The blue fescue needs very little maintenance and will look good with just a tidy to remove dead leaves in summer. Plants are long lived but can become a bit tired and may need replacing when they are around 5-7 years old.

ASPECT: Full sun
FOLIAGE INTEREST: Year round
HARDINESS: –10 to –15°C (14 to 5°F)

10–30 cm (4–12 in)

10–30 cm (4–12 in)

Deschampsia cespitosa

***Festuca glauca* 'Elijah Blue'**

HAKONECHLOA MACRA 'AUREOLA'

Golden hakonechloa

POACEAE

A deciduous grass that forms mounds of variegated bright golden-yellow and green striped leaves. It is an extremely attractive grass and I have included it because for a short while in early winter the striking summer foliage turns to shades of copper and orange.

Unfortunately, it dies back to the ground before the end of winter and, like all deciduous grasses, needs cutting back before the new growth appears in early spring.

ASPECT: Full sun, partial shade
FOLIAGE INTEREST: Year round
HARDINESS: −10 to −15°C (14 to 5°F)

30–50 cm (12–20 in)

30–50 cm (12–20 in)

MISCANTHUS NEPALENSIS

Himalayan fairy grass

POACEAE

Similar in many ways to *Miscanthus sinensis* but the pale yellow flowering heads are more open and delicate. As they mature through late autumn into winter the plumes of seedheads become more golden. This deciduous grass also offers interest for most of the year, its tall, linear green leaves turning bronze in winter, when a touch of frost adds to its appeal. It works well in a mixed border, where its open habit makes it easier for other tall perennials to grow around it, and looks particularly good with *Verbena bonariensis*.

ASPECT: Full sun, partial shade
INTEREST: Most of the year
HARDINESS: −10 to −15°C (14 to 5°F)

1–1.5 m (3–5 ft)

0.5–1 m (20–39 in)

At its best in early winter, when its copper and bronze tones are highlighted by the low sun

***Hakonechloa macra* 'Aureola'**

MISCANTHUS SINENSIS

Chinese silver grass

POACEAE

A beautiful architectural grass that has great winter appeal. In late summer or early autumn, the flowerheads form large silver or pale pink feathery plumes. They are held on tall stiff stems, well above the upright leaves, which are borne along the lower part of the stem and arch gracefully at the top. The leaves themselves turn straw-coloured in late autumn and throughout the winter.

The species is quite a large grass and probably not suited to the smallest gardens, as it creates more impact when planted in a group, but a couple in the corner of a bed as a backdrop to other plants still look amazing, especially when enhanced by a touch of frost.

Alternatively, choose one of smaller cultivars of this popular grass. A good medium-sized one that would work well in most gardens is 'Flamingo', with bright pink plumes that reaches around 1.5–2 m/5–6 ft. Or, for an even shorter selection, try 'Cindy', which is 1–1.5 m/3–5 ft tall.

Miscanthus needs to be cut back in late winter or early spring before the new growth appears.

ASPECT: Full sun, partial shade
INTEREST: Year round
HARDINESS: −10 to −15°C (14 to 5°F)

2–3 m (6–10 ft)

1–1.5 m (3–5 ft)

NASSELLA TENUISSIMA

Mexican feather grass

POACEAE

Until recently, this grass was known as *Stipa tenuissima* and will still be found under this name in gardens centres and online nurseries.

It is such a versatile little deciduous grass and probably better known for

Miscanthus sinensis 'Flamingo'

growing in a summer drought-tolerant garden, admired for its ability to hang on in the hottest of seasons. However, it is also very useful in winter, when its wispy leaves turn buff-coloured and look particularly attractive when covered with frost, or after a light snowfall.

Nassella tenuissima needs to be cut back in late winter or early spring before the new growth appears.

ASPECT: Full sun
INTEREST: Year round
HARDINESS: –10 to –15°C (14 to 5°F)

0.5–1 m (20–39 in)

10–50 cm (4–20 in)

Miscanthus nepalensis

Nassella tenuissima

Ophiopogon planiscapus 'Black Beard'

Pennisetum alopecuroides 'Hameln'

OPHIOPOGON PLANISCAPUS 'BLACK BEARD'

Black mondo grass

ASPARAGACEAE

Although this plant looks like and has characteristics of a grass, it actually belongs to the same family as asparagus and lily-of-the-valley (*Convallaria majalis*)! All these plants are distinguished by their bell-shaped flowers, which in this evergreen perennial are purple-flushed white and appear in summer, followed by shiny, almost black berries.

However, it is grown mostly for its black, grass-like, arching leaves that form spreading clumps. It creates a dark, textured ground cover that is ideal for combining with contrasting colours, particularly the white flowers of snowdrops, or white-stemmed birches.

Ophiopogon planiscapus 'Nigrescens' is a similar dark-leaved cultivar and widely available.

Relatively slow spreading, black mondo may need thinning out from time to time if plantings become congested.

ASPECT: Full sun, partial shade
FOLIAGE INTEREST: Year round
HARDINESS: –10 to –15°C (14 to 5°F)

10–30 cm (4–12 in)

10–30 cm (4–12 in)

PENNISETUM ALOPECUROIDES 'HAMELN'

Chinese fountain grass

POACEAE

This compact fountain grass produces short clumps of green linear leaves that turn gold in winter and flower spikes in summer that resemble small silvery bottle brushes, which persist throughout the colder months.

Pennisetum needs to be cut back in late winter or early spring before the new growth appears.

ASPECT: Full sun
FOLIAGE INTEREST: Year round
HARDINESS: –10 to –15°C (14 to 5°F)

0.5–1 m (20–39 in)

0.5–1 m (20–39 in)

PENNISETUM 'FAIRY TAILS'

Fountain grass

POACEAE

A fairly compact form of fountain grass that during the summer produces decorative flower spikes which look like long, silvery-white animal tails held above and among the dark green foliage. The flower spikes arch outwards from the

clump, resembling water spraying from a fountain, hence its common name.

During late autumn and early winter, the foliage turns golden-yellow and usually remains an attractive feature throughout the season.

ASPECT: Full sun
FOLIAGE INTEREST: Year round
HARDINESS: –10 to –15°C (14 to 5°F)

0.5–1 m (20–39 in)

0.5–1 m (20–39 in)

UNCINIA RUBRA

Red hook sedge

CYPERACEAE

Uncinia rubra, as its name suggests, forms small clumps of red or reddish-brown foliage. It is evergreen but may need some protection in colder, more exposed areas. The beauty of this grass-like plant is that there isn't another that offers these striking tones, but it needs a sunny position to produce the most intense colour. Try growing it in a sheltered spot in a container with other dwarf winter-flowering bulbs, perennials, and sub-shrubs.

Uncinia egmontiana is the orange hook sedge and, like the species above, it is not fully hardy but will do well in a container with other winter-flowering or interesting plants in a sheltered spot.

ASPECT: Full sun, partial shade
FOLIAGE INTEREST: Year round
HARDINESS: 1 to –5°C (34 to 23°F)

10–50 cm (4–20 in)

10–50 cm (4–20 in)

***Pennisetum alopecuroides* 'Fairy Tails'**

Bamboo

There are many different genera of bamboo, a few of which make ideal plants for including in a winter planting scheme. Bamboos are in the grass family, Poaceae.

Not only are they evergreen and very hardy, some also have striking, brightly coloured or unusually shaped stems, known as culms or canes. Take care when selecting bamboos as those that spread or run can be very invasive, while those that form clumps are generally better behaved and more suitable for garden use. I have therefore chosen two clump-forming types here, as some of the aggressive running bamboos can be a nuisance, not only in the garden they are planted in, but also in neighbouring plots. However, even bamboos that form clumps do sometimes spread beyond their allotted space, so restrict their growth with a non-perishable barrier. The best method is to use a large plastic pot, at least 60 cm/ 2 ft deep, with the bottom removed to allow good drainage. Place the pot in the ground, with the rim just above the soil level. You then plant the bamboo into the pot and it will prevent any runners that are produced from escaping into the surrounding soil. However, it is not 100 per cent foolproof, so keep an eye out for any shoots outside of the potted barrier area.

Bamboo is technically a woody grass but it does not need cutting back each year like most perennial grasses, just the occasional removal of the oldest canes.

PHYLLOSTACHYS AUREA
Fishing pole bamboo

POACEAE

The genus *Phyllostachys* includes tall, architectural plants that look fantastic as individual specimens, but work equally planted among other shrubs, or as a screen. They also make a very attractive containerised plant, if kept well-watered and are divided regularly or repotted in a larger container every few years.

OPPOSITE
***Phyllostachys aureosulcata* 'Aureocaulis'**

Black bamboo

The tall stems of the species *Phyllostachys aurea* are pale green when young and covered in small clusters of short, lance-shaped leaves, held on twiggy growth, which flutter attractively and rustle in the breeze. As the canes mature, they become golden yellow. For the best colour, grow this bamboo in full sun and cut some of the lower twiggy growth off each cane, which will reveal more of the stem and, by exposing it to more light, also help to intensify the golden hue. New canes are produced early in the year and plants must be kept well-watered at this time. Any old or damaged canes that need removing should be cut down as low as possible to the ground. These can then have a second life in the garden as plant supports.

The common name, the fishing pole bamboo, reminds me that when I started fishing as a boy in the 1960s my first rods were made from split canes.

Phyllostachys nigra – black bamboo – is very similar to *Phyllostachys aurea* in all but colour. As the name suggests, its canes become black with age. Again, choose a sunny spot for the best colour. Planted together, the different coloured stems contrast and complement each other.

ASPECT: Full sun, partial shade
INTEREST: Year round
HARDINESS: –10 to –15°C (14 to 5°F)

4–8 m (12–25 ft)

1.5–3 m (5–10 ft)

PHYLLOSTACHYS AUREOSULCATA F. AUREOCAULIS
Golden groove bamboo

POACEAE

The large and impressive bright yellow canes of this bamboo are definitely among the best of the yellow varieties. The stems are often reddish when young and streaked with green as they mature – the green area is usually more prominent towards the base of the canes. This striking bamboo also has a habit of producing a few bendy stems that zigzag at angles from the nodes, giving the plant a distinctive appearance.

The yellow canes contrast well with the darker green, lance-shaped leaves. This architectural feature plant will light up a shadier area of the garden, as the yellow stems do not need as much light as other colourful bamboos, but it is particularly beautiful when illuminated by bright winter sunshine.

It may also be found under the name *Phyllostachys aureosulcata* 'Aureocaulis'.

ASPECT: Full sun, partial shade
INTEREST: Year round
HARDINESS: –10 to –15°C (14 to 5°F)

4–8 m (12–25 ft)

2–4 m (6–12 ft)

Perennials

Ajuga reptans

AJUGA REPTANS

Bugle

LAMIACEAE

A great little ground-cover plant which produces runners that spread across the surface and then take root to produce new plants. Bugle is native to Great Britain and Ireland, where it is found growing in rough grassland and semi-shaded areas of woodland, and it is in these sites that it will do best in a garden.

In its wild form, the evergreen leaves are a purplish-green, but it is the many cultivars that offer the most interest in a winter garden. Foliage colours range from dark green to bronze and dark purple, while the variegated forms, such as *Ajuga reptans* **'Burgundy Glow'**, are decorated in a mixture of hues. The dark leaves of *Ajuga reptans* **'Black Scallop'** or **'Atropurpurea'** look particularly good in a mixed planting with snowdrops, the foliage developing a darker colour when grown in open, sunny positions.

Ajuga also produces flowers (mainly blue) in early spring and summer that are attractive to bees and butterflies.

If carpets become too large, remove the runners and replant them in other areas of the garden, or in containers.

ASPECT: Full sun, partial shade
INTEREST: Year round
HARDINESS: –10 to –15°C (14 to 5°F)

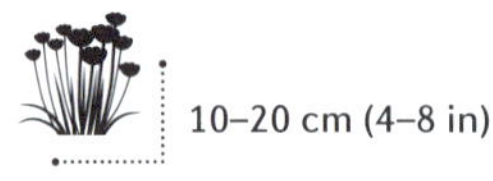

10–20 cm (4–8 in)

20–80 cm (8–32 in)

BERGENIA PURPURASCENS

Purple bergenia

SAXIFRAGACEAE

Commonly known as elephant's ears, *Bergenia* is a little like bugle on steroids and also has many garden-worthy cultivars. The rich dark purple-leaved types are favoured most for use in winter planting schemes, and this species is as good for colour as many of the cultivars. For most of the year the large, leathery, oval leaves are dark green, with some red tints on their undersides, but during the winter months they turn a wonderful beetroot red. Plants grown in sun produce a better, deeper colour than those in shade.

Bergenias are clump-forming and can be divided in spring after flowering or in autumn if they spread too far. Attractive pink flowers are produced on tall spikes in early spring.

ASPECT: Full sun, partial shade
FOLIAGE INTEREST: Year round
HARDINESS: –10 to –15°C (14 to 5°F)

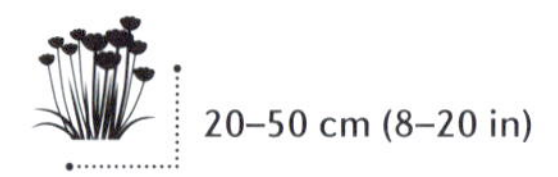

20–50 cm (8–20 in)

20–50 cm (8–20 in)

Bergenia purpurascens

Chrysosplenium macrophyllum

CHRYSOSPLENIUM MACROPHYLLUM

Giant golden saxifrage

SAXIFRAGACEAE

Not commonly seen in gardens, the giant golden saxifrage is an attractive, ground-hugging and slow-spreading evergreen perennial that deserves to be grown more widely. In late winter and early spring it produces clusters of white flowers with prominent pink stamens on stiff reddish stems above dark green foliage. Each small cluster of flowers is surrounded by larger silvery bracts with pink margins which remain long after the blooms have faded, extending the plant's period of interest.

During the colder months, its fleshy dark green foliage becomes flushed with reddish-brown, adding yet more colour.

Spreading by long runners, this perennial will slowly increase in size and is an ideal plant for a shady area where it will thrive in the cool, moist conditions. In drier positions, it is less likely to spread.

ASPECT: Partial shade

FLOWERING: February – March

HARDINESS: –10 to –15°C (14 to 5°F)

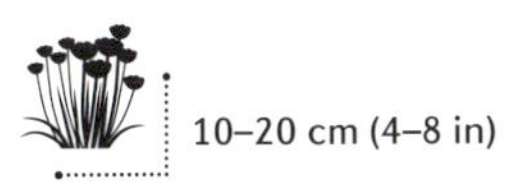

10–20 cm (4–8 in)

10–50 cm (4–20 in)

EUPHORBIA CHARACIAS SUBSP. WULFENII

Mediterranean spurge

EUPHORBIACEAE

This tall architectural plant has many attributes that make it ideal for a winter garden.

It is an evergreen that produces large clumps of upright stems adorned with bluish-green foliage that are enhanced during frosty periods, when they glisten in the early morning winter sun.

Just as winter is coming to an end, the leafy stems are topped by huge heads of small lime-green flowers, each made up of dozens of individual bell-shaped blooms with dark purple glands in the centre.

The flower spikes are biennial, so after they have flowered in their second year, they should be cut right back to the ground, but I do this in the second spring so that the dried flowers can also add winter interest.

Wear thick gloves when pruning any part of this plant as it produces a milky sap that can irritate the skin.

ASPECT: Full sun

FLOWERING: February – May

HARDINESS: –5 to –10°C (23 to 14°F)

0.8–1 m (32–39 in)

0-5–1 m (20–39 in)

Such a versatile plant year-round. In late winter it produces its large flower heads, which continue to look good through spring into early summer

Euphorbia characias subsp. wulfenii

Hellebores

Of all the winter-flowering perennials, hellebores are probably the most useful. While other plants are waiting for warmer and longer days to perform, hellebores are in full bloom in winter, brightening up cold, bleak days and thriving through frost and snow. In fact, these conditions actually help to enhance their beauty.

They come in a wide range of colours, from the purest of whites to the deepest purples – some are almost black – and many colours in between. Some are also multi-coloured with blotches, stripes, and speckling. Flowers can be either single or double forms, and the flowering season runs from early to late winter and into spring. The sterile hybrids flower for the longest.

Typically woodland edge plants, hellebores are perfect for growing in the shadier parts of a garden where they help to create naturalistic and informal planting designs.

Helleborus argutifolius

HELLEBORUS ARGUTIFOLIUS
Corsican hellebore

RANUNCULACEAE

The mid-green, shiny, prickly edged leaves of the Corsican hellebore are a feature in themselves before and after flowering, and provide a backdrop to the clusters of lime green, bowl-shaped blooms with pale yellow stamens, which appear in winter.

The flowers are borne on thick, sturdy stems and last for many weeks above the contrasting darker leaves. The blooms eventually drop their stamens as they mature, which changes their appearance, but they continue to perform for some time afterwards. This plant prefers dappled shade but will do equally well in a sunny border.

ASPECT: Full sun, partial shade
FLOWERING: January – April
HARDINESS: –5 to –10°C (23 to 14°F)

50–75 cm (20–30 in)

0.5–1 m (20–39 in)

HELLEBORUS X HYBRIDUS
Hybrid Lenten rose

RANUNCULACEAE

This evergreen hellebore is extremely variable, and if bought in leaf it will be potluck as to the colour of the flowers the plant will produce. The blooms range from white and pink to mauve and deep purple, while many have patterned petals that include all these colours. So if you are planning a particular colour scheme, it's best to buy plants in flower.

The Hybrid Lenten rose will produce seedlings freely, which can be removed easily if not wanted, or used to bulk up a planting with many interesting colours and combinations – seedlings collected and potted up while still young develop well and can then be transplanted elsewhere. Cutting back plants' old evergreen foliage as the new flower stems and leaves are emerging will not only show off the blooms to best effect, but also remove any suffering from the fungal disease, hellebore leaf spot, preventing it from spreading.

ASPECT: Partial shade
FLOWERING: January – March
HARDINESS: –10 to –15°C (14 to 5°F)

30–50 cm (12–20 in)

30–50 cm (12–20 in)

HELLEBORUS NIGER
Christmas rose

RANUNCULACEAE

The Latin species name *niger* (meaning black) might suggest that the Christmas rose is dark in colour – one of its other common names is the black hellebore – but it is actually the plant's roots that are black, while the flowers are usually the purest white or, occasionally, flushed with pink.

The large, open, bowl-shaped flowers with golden stamens in the centre are produced singly on short, sturdy stems just above the dark green, leathery, evergreen leaves. Unlike many hellebore flowers, those of the Christmas rose do not droop but face outwards and contrast well with the dark foliage. The plant's compact size also makes it ideal for growing in containers as part of a mixed winter display.

ASPECT: Partial shade
FLOWERING: January – March
HARDINESS: –10 to –15°C (14 to 5°F)

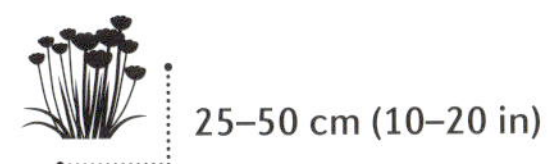

25–50 cm (10–20 in)

25–50 cm (10–20 in)

HELLEBORUS
'Anna's Red'

RANUNCULACEAE

This pretty hellebore is part of a relatively new hellebore hybrid group known as the Rodney Davy Marbled Group, noted for their distinctive marbled foliage. This particular variety is named after the plantswoman Anna Pavord.

An interspecies hybrid, it produces a profusion of larger than average flowers, each bloom forward-facing rather than drooping. The foliage is marbled with silver veining, almost like flashes of lightning. Like all hellebores, it works well planted among winter-flowering bulbs.

There are also pink (**'Penny's Pink'**) and white (**'Moondance'**) cultivars in the group that have the marbled foliage and large forward-facing flowers.

ASPECT: Full sun, partial shade
FLOWERING: December – March
HARDINESS: –10 to –15°C (14 to 5°F)

30–60 cm (12–24 in)

30–60 cm (12–24 in)

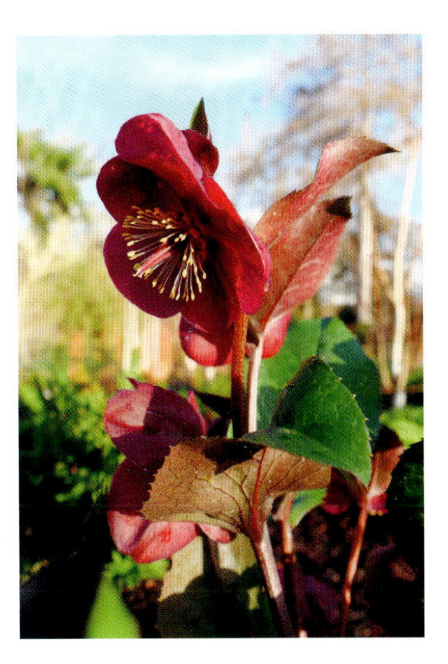

***Helleborus* 'Anna's Red'**

HELLEBORUS
'Walberton's Rosemary'

RANUNCULACEAE

This hybrid cross between *H. niger* and *H.* x *hybridus* is one of many bred at Walberton's nursery in West Sussex in the UK.

It's a sterile hybrid that produces lots of outward-facing pink flowers with paler veining, held on tall stems, which contrast well with the deep green leaves. A showy plant, it's a joy when the stunning flowers appear in midwinter. It will thrive when planted in a woodland-type setting, but it does equally well in a sunny bed or border among other plants. Remove old foliage in autumn before the new flower stems emerge to reduce the risk of fungal disease and show off the new flowers.

ASPECT: Full sun, partial shade
FLOWERING: December – March
HARDINESS: –10 to –15°C (14 to 5°F)

30–60 cm (12–24 in)

30–60 cm (12–24 in)

Helleborus* x *hybridus

Helleborus niger

Pachysandra terminalis 'Variegata'

HEUCHERA 'PURPLE PETTICOATS'

Coral bells

SAXIFRAGACEAE

There are many different cultivars of heuchera that offer a variety of colours and leaf shapes. The popular 'Purple Petticoats' has dark, maple-like foliage, marbled metallic bronze–purple on top and paler pink beneath, while other cultivars and varieties range from pale lime green, pink and red to the darkest of purples, some appearing almost black, and variegated forms.

Heucheras are all evergreen, or semi-evergreen, small clump-forming perennials that can be used as ground cover, particularly in dappled shade, or in pots and containers as part of a mixed winter scheme. In frosty conditions, their leaf margins will glisten and sparkle in the winter sunlight.

Clumps can be lifted and divided every few years in late spring to maintain their vigour.

ASPECT: Full sun, partial shade
FOLIAGE INTEREST: Year round
HARDINESS: –10 to –15°C (14 to 5°F)

10-50 cm (4–20 in)

10–50 cm (4–20 in)

PACHYSANDRA TERMINALIS 'VARIEGATA'

Variegated Japanese spurge

BUXACEAE

This tough spreading evergreen is technically a subshrub, but I have listed it under perennials as this is how it is most commonly treated. It is a great plant for using as ground cover in a shady spot and will even thrive in full shade. Its pale green leaves have cream margins which help to brighten up areas beneath and around shrubs. Spreading via fleshy rhizomes, it will eventually form a dense blanket that will suppress weeds.

Greenish-white unisexual flowers are produced in late winter or early spring on short spikes above the foliage – the male flowers have prominent white stamens.

There is also an all-green cultivar *Pachysandra* **'Green Carpet'**.

This low-maintenance plant needs very little attention, other than lifting and removing any creeping stems that outgrow their allotted space. These rooted runners can be replanted elsewhere in the garden.

ASPECT: Partial shade, full shade
FOLIAGE INTEREST: Year round
HARDINESS: –10 to –15°C (14 to 5°F)

10–30 cm (4–12 in)

10–50 cm (4–20 in)

PRIMULA VULGARIS

Primrose

PRIMULACEAE

One of the many joys of the English countryside is the appearance of wild primroses in late winter and early spring. Their lemon-yellow (primrose-coloured) flowers sit among a rosette of wrinkled green leaves and are a great addition to any area of a garden, along with snowdrops and winter aconites.

The plant's name derives from the Latin *prima rosa*, meaning 'first rose' of the year, because the flowers look similar to a single rose, though it is not in the rose family. Each flower has a darker yellow centre produced by spots at the base of each petal.

There are lots of cultivars in various colours, but for me the 'original' yellow is the best.

This primrose will spread naturally and self-sown seedlings are easily removed if not wanted.

ASPECT: Full sun, partial shade
FLOWERING: February – April
HARDINESS: –10 to –15°C (14 to 5°F)

5–10 cm (2–4 in)

10 –20 cm (4–8 in)

PULMONARIA RUBRA

Red lungwort

BORAGINACEAE

One of the first of the lungworts to flower, this pretty, medium-sized perennial makes an excellent ground-cover plant in shady areas, where it will produce clusters of deep pink to red, funnel-shaped flowers, above green and white mottled foliage. There are not many other winter-flowering plants that are this colour, and it makes an attractive contrast to hellebores and early bulbs.

Like most *Pulmonaria*, it is semi-evergreen, but rarely loses its leaves, except in the coldest years in exposed areas open to the worst of the winter weather.

It is a great plant for attracting early flying bees that venture out on warmer winter days, collecting pollen and feeding on nectar.

The common name lungwort refers to its mottled leaves, which supposedly resemble diseased lungs, which are also referenced in its Latin name *Pulmonaria*.

The attractive pink flowers always entice early flying bumblebees to feed on their nectar

Pulmonaria rubra

Remove the old leaves after flowering to keep it tidy.

ASPECT: Partial shade, full shade
FLOWERING: February – April
HARDINESS: –15 to –20°C (5 to– 4°F)

10-30 cm (4–12 in)

10–30 cm (4–12 in)

Primula vulgaris

VINCA MINOR 'ALBA AUREOVARIEGATA'

Lesser variegated periwinkle

APOCYNACEAE

Pure white star-shaped flowers partnered with glossy leaves featuring irregular yellow variegations make this little evergreen perennial an attractive flowering ground-cover plant. Tolerating full sun to almost full shade, it is ideal for difficult situations such as sprawling over banks and trailing down slopes, or brightening up a spot under an evergreen shrub. It also works well in containers as a trailing plant.

Spreading by trailing stems, which form roots at their nodes, in ideal conditions it can quickly cover a wide area, but can be easily controlled by removing the excess growth before it takes root.

ASPECT: Full sun, partial shade
FLOWERING: February – May
HARDINESS: –10 to –15°C (14 to 5°F)

10–20 cm (4–8 in)

1–1.5 m (3–5 ft)

Shrubs

ABELIOPHYLLUM DISTICHUM

White forsythia

OLEACEAE

A deciduous shrub that's related to forsythia and looks very similar to it, hence the common name, this species of *Abeliophyllum* has a sprawling habit and can be grown as a free-standing specimen in a mixed shrub border or trained against a wall or fence. Before the leaves appear in spring, its long, arching stems produce small clusters of four-petalled white or pink winter flowers with a delightful almond scent. To enjoy their delicate fragrance to the full, take a few cut stems into the house and display them with other winter flowers.

The pale pink forms belong to the Roseum Group, and while they are not as common as the white one, the blooms emit an equally delicate scent.

Prune the long stems after flowering to keep the plant bushy and prevent it from become straggly.

ASPECT: Full sun, partial shade
FLOWERING: January – February
HARDINESS: –10 to –15°C (14 to 5°F)

1.5–2 m (5–6 ft)

1.5–2 m (5–6 ft)

AUCUBA JAPONICA

Japanese laurel

GARRYACEAE

The species *Aucuba japonica* is a large evergreen shrub that was frequently used in Victorian times as a screening plant and may not be everyone's first choice for a winter garden planting, particularly where space is limited. However, this tough shrub is extremely tolerant of a wide range of conditions, including dry soil and shade, and it can be very useful if you can fit it in.

Abeliophyllum distichum

Aucuba japonica

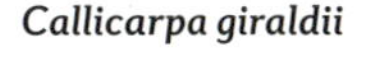

Callicarpa giraldii

It is also tolerant of salt-laden wind, and therefore a good plant for coastal gardens too.

The glossy foliage is dark green and through late autumn and winter female plants produce large bright red ovoid berries which stand out well against the leaves, but you will need a male plant nearby for the fruit to form.

This shrub requires very little pruning but can be cut back in early spring if plants become too large or congested.

ASPECT: Partial shade, full shade
FOLIAGE INTEREST: Year round
HARDINESS: −10 to −15°C (14 to 5°F)

1.52 m (56 ft)

1.5–2.5 m (5–6 ft)

AUCUBA JAPONICA 'VARIEGATA'

Variegated Japanese laurel

GARRYACEAE

This Japanese laurel is not everyone's cup of tea and people tend to either love it or hate it. But whatever your view, it is an amazingly versatile evergreen shrub that will tolerate almost any site and soil conditions, including dry shade, once established. This makes it very useful as a screening planting, particularly in a dark corner and under trees where its yellow variegated markings will bring some flecks of colour.

Aucuba japonica 'Variegata'

It can also be used as a backdrop for brighter-coloured plants, and the female clone will produce red berries in late autumn if a male clone is nearby.

This shrub requires very little pruning but can be cut back in early spring if plants become too large or congested.

ASPECT: Partial shade, full shade
FOLIAGE INTEREST: Year round
HARDINESS: −10 to −15°C (14 to 5°F)

1.5–2 m (5–6 ft)

1.5–2.5 m (5–6 ft)

CALLICARPA GIRALDII

Beautyberry

LAMIACEAE

Following many name changes, this tall deciduous shrub is now most commonly sold as the cultivar *Callicarpa bodinieri* var. *giraldii* 'Profusion'.

The violet–purple berries that are borne in small clusters along its stems are among the showiest of all the winter berries. They begin to appear in late autumn just as the leaves are turning shades of pink and purple, which contrast well with the almost iridescent fruits. The berries persist throughout the months that follow, decorating the bare stems and providing a show long after the foliage has fallen.

Prune in early or mid-spring, removing any stems that have become old and woody to the ground.

ASPECT: Full sun, partial shade
FRUIT INTEREST: October – January
HARDINESS: −10 to −15°C (14 to 5°F)

2.5–3 m (8–10 ft)

2–2.5m (6–8 ft)

Camellias

Native to South East Asia, camellias have been in and out of fashion in Europe since the early eighteenth century. Originally thought to be tender and difficult to grow, these evergreen shrubs were kept in glasshouses until later when it was discovered that they are, in fact, among the hardiest winter-flowering shrubs, blooming throughout the coldest months of the year, with many able to withstand temperatures as low as –10°C (14°F). However, prolonged periods of colder weather may damage the flower buds.

Native to woodlands, these plants do best in partial shade, ideally under trees, and in an acid soil, but many will also tolerate soils that are neutral (pH7). In gardens with alkaline soil, they can be successfully grown in an acidic (ericaceous) compost in large containers. Good drainage is also essential.

There are many species of camellias, and I have included those that are the most garden-worthy, together with just a few of their cultivars and hybrids, but look out for new varieties, which are being bred all the time.

Camellias require very little pruning, but they can be cut back in spring after flowering, if needed, to remove unwanted stems or new growth that has been damaged by late frosts.

Camellia japonica

CAMELLIA JAPONICA

Common camellia

THEACEAE

This species of camelia has produced many cultivars, but in the wild the flowers usually have six or seven petals and are either red or white. However, the cultivars that have been produced around the world come in a much wider range of colours, from white to the darkest red, as well as bi-coloured blooms with blotches, specks or stripes. The flowers can also be single, semi-double, or double, with numerous petals.

Camellia japonica 'Lady Vansittart'

Apart from the wide choice of flower forms, the blooms appear over a long period, some lasting up to three or four months. The evergreen foliage makes them a good backdrop, too, highlighting paler plants in front, when included in a shady mixed border.

'Devonia' produces pure white, cup-shaped single flowers, with a large boss of golden stamens. It is quite vigorous with an erect habit. Height to 4 m/13 ft.

'Lady Vansittart' produces semi-double flowers that are unstable, which means that the flowers can look very different from each other, with plain white, pink, and red blooms and those with striped or blotched petals all appearing on the same plant. Height to 4 m/13 ft.

'Elegans' has large rose-pink flowers about 10 cm/4 in across. It is an anemone form, and the flowers have a mass of central petaloids (petal-like structures), often with white markings on them. This camellia was introduced from China in the mid-nineteenth century and it is still a popular cultivar today, producing a profusion of blooms over a long period from late winter to mid-spring. The flowers stand out well against the glossy, evergreen, dark green leaves. Height to 3 m/10 ft.

ASPECT: Partial shade
FLOWERING: February – April
HARDINESS: –5 to –10°C (23 to 14°F)

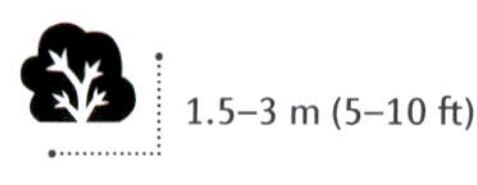

CAMELLIA SASANQUA

Sasanqua camellia

THEACEAE

This species and its cultivars start flowering in late autumn and continue to bloom on both sides of Christmas and into the new year, some even keeping up the show until early spring. More tolerant of a variety of conditions than *Camellia japonica*, they will grow in sunnier sites, and neutral, less acidic soils.

They are also a good choice for a small courtyard garden, growing well in containers and as a wall-trained shrub. Their single flowers are white, or occasionally rose-pink, with cultivars in many colours and forms, from single to fully double.

'Mine-no-yuki' flowers in November and December, producing large, scented, white double flowers, which can be a bit variable in form. It also has a graceful, pendulous habit. Height to 3 m/10 ft.

'Peach Blossom' has beautiful pale pink, semi-double flowers with rounded petals. Later-flowering than most sasanquas, with flowers appearing in late winter, usually in February. It forms a medium-sized shrub. Height to 3–4 m/10–13 ft.

'Red' is listed as autumn-flowering but will bloom from late autumn and right through December, often into January. It produces semi-double pinkish-red flowers with a central boss of showy yellow stamens. Height 3 m/10 ft.

ASPECT: Partial shade

FLOWERING: November – March

HARDINESS: −5 to −10°C (23 to 14°F)

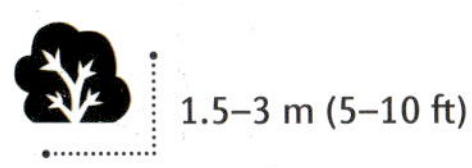

1.5–3 m (5–10 ft)

1–2 m (3–6 ft)

Camellia sasanqua

Williamsii hybrid camellias

These are hybrids produced from crossing *Camellia saluensis* and *Camellia japonica* (and their cultivars and varieties) that were originally bred in the early twentieth century by John Charles Williams of Caerhays Castle in Cornwall. The results are a range of camellias that get their hardiness from their *japonica* parent and long flowering period from the *saluensis* species.

'Bow Bells' is an upright shrub, with soft pink trumpet-shaped flowers that are produced over a long period. Height to 3–4 m/10–13 ft.

'Jury's Yellow' has creamy-yellow anemone-form flowers, with cream outer petals and numerous darker yellow petaloids (petal-like structures) in the centre. A small to medium-sized shrub. Height to 1.5–2 m/5–6 ft.

'St Ewe' makes a bushy shrub, with single deep pink flowers, bell-shaped at first, then opening wider as they mature. Height to 4 m/13 ft.

ASPECT: Full sun, partial shade
FLOWERING: January – March
HARDINESS: –5 to –10°C (23 to 14°F)

4 m (13 ft)

2 m (6 ft)

They have a long flowering period and are frequently visited by bees early in winter

Camellia x *williamsii* 'Bow Bells'

Camellia x *williamsii* 'Jury's Yellow'

Camellia x *williamsii* 'St Ewe'

Chaenomeles

Chaenomeles, also known as the ornamental quinces, are small to medium-sized deciduous shrubs that can be trained on walls and fences or grown as free-standing specimens. When choosing a quince, opt for one that flowers on bare stems before the leaves unfurl, since the blooms of some of the later varieties, particularly those with small single flowers, may be hidden by the foliage and have less impact.

Trained and grown as a wall shrub the flowers, which appear along bare stems, can be seen at their best

***Chaenomeles speciosa* 'Nivalis'**

CHAENOMELES JAPONICA
– Japanese quince – ROSACEAE

This species tends to be lower-growing than others in this genus, with mainly red–orange single flowers and thorny stems, but its cultivars are more varied and come in pinks, whites, and mixed colours. It will tolerate most soils and produces rounded fruits in autumn.

'Cido' has clusters of small-cupped, orange–red flowers borne along bare, thornless branches. This Latvian cultivar has the added bonus of producing large yellowish fruits in autumn with exceptional flavour, ideal for making jams and jellies. Height to 1.5 m/5 ft.

'Orange Beauty' is a multi-stemmed, low-growing shrub that produces showy orange cup-shaped flowers along its branches before the dark green shiny leaves are fully out. Height to 1 m/3 ft.

'Sargentii' has clusters of scarlet–orange cup-shaped single flowers along its bare branches in late winter. The last of the flowers nestle between glossy dark green, rounded leaves Height to 1.5 m/5 ft.

CHAENOMELES SPECIOSA
– Chinese quince - ROSACEAE

Growing up to 3 m/10ft tall, this large, rounded, deciduous shrub produces a tangle of spiny branches covered in clusters of wide bowl-shaped red flowers. The main flowering period is late winter, but it often has a second lighter flush of blooms in early summer when in leaf. Orangy-coloured apple-shaped fruits appear in the autumn.

'Madame Butterfly' is an outstanding ornamental quince and very showy. Its salmon-pink flowers are edged and streaked with white, the patterns running down each petal towards the yellow centre. Definitely a talking point. Height to 1.5 m/5 ft.

***Chaenomeles speciosa* 'Madame Butterfly'**

'Nivalis' in Latin means snowy, or snow-like, and the pure white flowers give the impression that this small shrub is covered in snow when they appear on the bare, spiny branches from late winter. Height to 2.5 m/8 ft.

'Kinshiden' is a particular favourite of mine. It produces clusters of pretty lime-green double flowers that deepen in colour in the centre. This shrub has a spreading habit and is fairly slow growing. Height to 1.5 m/5ft.

These quinces all require very little, if any, pruning – the crossing branches are part of their habit and can be left to grow naturally. Wear thick gloves to cut the prickly plants and remove any dead or diseased branches as they appear. If training against a wall, just restrict all growth coming away from the wall to two or three buds, unless the flexible stems are required to cover more of the wall or to fill gaps.

CHAENOMELES X SUPERBA
– Japanese quince – ROSACEAE

Produced from crossing *Camellia japonica* and *Camellia speciosa*, this hybrid and its cultivars are vigorous,

Chaenomeles speciosa 'Kinshiden'

medium-sized, spreading shrubs.

'Cameo' is a double-flowered peach-pink cultivar, its spiny stems creating a neat, rounded habit. It begins to flower before the leaves are produced, but the blooms still sit well against the backdrop of foliage when it appears. Height to 1 m/3 ft.

'Crimson and Gold' is one of the best known and most striking of all flowering quinces. The bright red or crimson flowers are set off by golden anthers in the centre. It looks great trained against a wall, or used as a free-standing shrub, when its tangle of branches are adorned with flowers. Height to 1.5 m/5 ft.

'Lemon and Lime', as the name suggests, has lemon- and lime-coloured flowers. The petals of the cup-shaped blooms are pale yellow, darkening to a lime green in their centre, and, like many of these hybrids, they are borne along spiny tangled branches. Height to 1.5 m/5 ft.

This hybrid's vigour may need to be controlled by thinning out and pruning the stems in midsummer.

ASPECT: Full sun, partial shade
FLOWERING: February – April
HARDINESS: −10 to −15°C (14 to 5°F)

1–3 m (3–10 ft)

1.5–3 m (5–10 ft)

Chaenomeles x *superba* 'Cameo'

Chaenomeles x *superba* 'Crimson and Gold'

CHIMONANTHUS PRAECOX

Wintersweet

CALYCANTHACEAE

Wintersweet is an apt name for this pretty deciduous shrub. Its pale cream or yellow flowers are highly scented and fill the air over a wide area around the plant with their sweet aroma.

The waxy-looking, bowl-shaped flowers hang down from the bare branches, either singularly or in small clusters, on very short stems. Flower colour is quite variable, usually creamy-white to pale yellow, with darker purplish smaller petals in the centre.

Chimonanthus praecox 'Grandiflorus', as the name suggests, has larger flowers than the species. The blooms are yellow with a red centre (although this is variable).

Chimonanthus praecox 'Luteus' flowers slightly later than most wintersweets, its yellow blooms opening in February.

These shrubs are good choices for training against a sunny wall, and the stems can also be cut and placed in a vase indoors to fill a room with their wonderful scent.

Wintersweet should only be pruned once the shrub has produced a mature branch system. The fragrant flowers are produced on the current year's growth, so they should be pruned straight after blooming to allow a full season of new wood to be produced.

ASPECT: Full sun
FLOWERING: December – February
HARDINESS: –10 to –15°C (14 to 5°F)

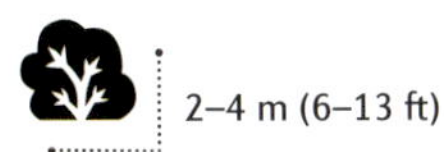

2–4 m (6–13 ft)

1.5–2.5 m (5–8 ft)

Chimonanthus praecox

***Chimonanthus praecox* 'Grandiflorus'**

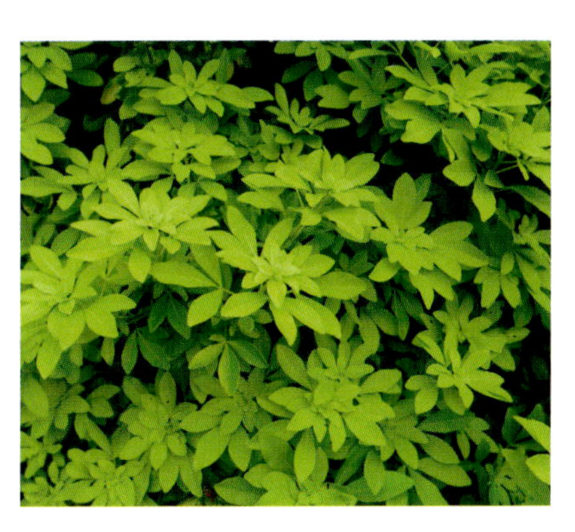

***Choisya ternata* 'Sundance'**

CHOISYA TERNATA 'SUNDANCE'

Mexican orange blossom

RUTACEAE

A medium-sized evergreen shrub with glossy, bright golden-yellow fragrant leaves, this dome-shaped Mexican orange blossom works well in a mixed border. Grow it in full sun to enhance the colourful foliage, which is brightest when young and fades to yellow–green with age.

The fragrance from the leaves is only apparent when the leaves are crushed. The common name Mexican orange blossom refers to the scented flowers that appear in late spring and are reminiscent of orange tree blooms.

This plant that will grow in most sites and soils, and, as well as being a great shrub for a winter garden, it is also very drought tolerant once established.

It requires little maintenance, apart from a light prune after flowering in early summer, if needed, to reduce its size.

ASPECT: Full sun, partial shade
FOLIAGE INTEREST: Year round
FLOWERING: Main period: April – June
HARDINESS: –5 to –10°C (23 to 14°F)

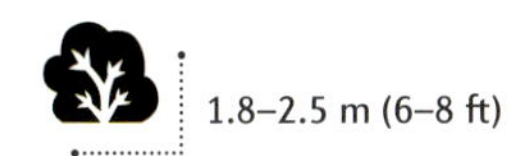

1.8–2.5 m (6–8 ft)

1.5–2.5 m (5–8 ft)

Kew's iconic Palm House framed by the glowing winter stems of *Cornus* 'Midwinter Fire'

CORNUS
Dogwoods
The different species of *Cornus* (known as dogwoods) and their cultivars are mainly grown for their vibrant winter stem colours, which range from yellow through to deep purple. The brighter and sunnier their position, the better the colour will be.

After one or two years, when new plants have established, they should be pruned hard each spring (March), just before they come into leaf, to a bud approximately 5 cm/2 in above ground level, to encourage bright new shoots the following winter.

***Cornus sanguinea* 'Midwinter Fire'**

CORNUS ALBA 'SIBIRICA'
– Siberian dogwood – CORNACEAE
The best known of the red-stemmed dogwoods, and very similar to the cultivar 'Westonbirt' (they may actually be one and the same thing), this deciduous suckering shrub forms large groups of tall, slender crimson stems in winter. Left unpruned, it will produce flat heads of creamy-white flowers in early summer followed by bluish-white berries. The species name *alba* means white and refers to the berries. The green foliage also turns red in autumn.

'Kesselringii' produces deep purple, almost black, stems and contrasts well with yellow-stemmed dogwoods, which will really stand out against this dark cultivar. Like 'Sibirica', it flowers in summer and produces white fruits and reddish-purple leaf colours in autumn.

CORNUS SANGUINEA
'MIDWINTER FIRE'
– Common dogwood – CORNACEAE
The cultivar 'Midwinter Fire' has been around for many years and its brilliant flame-coloured stems are orangey-yellow, turning redder towards the tips. It looks best planted en masse, when it resembles a flaming bonfire. It is not as vigorous as other cultivars, though, so do not cut back as hard or as often as *C. alba* cultivars. 'Winter Beauty' and 'Winter Flame' are both similar to 'Midwinter Fire'.

'Anny's Winter Orange' is the brightest of all the dogwoods. It is also stronger and more vigorous, and my personal favourite.

CORNUS SERICEA
– Red osier dogwood – CORNACEAE
Although this species of dogwood has red stems, the best of its cultivars are green and yellow, and commonly called yellow-twig dogwoods.

'Bud's Yellow' has bright yellow stems that look particularly eye-catching when planted with contrasting red dogwoods.

'Flaviramea' has yellow–green, or yellow–olive-coloured stems.

ASPECT: Full sun, partial shade
STEM INTEREST: December – March
HARDINESS: –10 to –15°C (14 to 5°F)

1–1.5 m (3–5 ft)

1.5–2 m (5–6 ft)

Cornus alba 'Sibirica'

Cornus sanguinea 'Anny's Winter Orange'

Cornus sericea 'Bud's Yellow'

Coronilla valentina subsp. *glauca*

CORONILLA VALENTINA SUBSP. *GLAUCA*

Glaucous scorpion vetch

FABACEAE

This amazing medium-sized, evergreen shrub is native to the Mediterranean but hardy in most winters in the UK, dealing well with temperatures down to about −10°C/14°F. It also produces clusters of fragrant, rich yellow, pea-like flowers throughout the year, with the main flush starting in April, but it always has some blooms around Christmas time too. The cultivar **'Citrina'** has paler lemon-yellow flowers.

Even though it doesn't produce flowers of the same quantity during the winter months compared to the summer, it will be visited by bees, particularly bumblebees, during mild spells.

Very little, if any, pruning is needed, apart from the removal of dead, diseased, and damaged stems.

ASPECT: Full sun, partial shade
FLOWERING: Year round; main period April– June
HARDINESS: −5 to −10°C (23 to 14°F)

2–3 m (6–10 ft)

1.5–2 m (5–6 ft)

Corylus avellana 'Contorta'

CORYLUS AVELLANA 'CONTORTA'

Corkscrew hazel

BETULACEAE

This large deciduous shrub is at its best during the winter months, when the twisted or contorted branches can be seen more clearly. During late winter it produces long golden-yellow male catkins that sway attractively in the breeze. It looks particularly good underplanted with snowdrops and hellebores and is relatively slow growing, which also makes it ideal for growing in a large container.
There is a purple-leaved cultivar **'Red Majestic'** which bears purplish-red catkins.

Apart from cutting out the odd bit of dead wood, the only pruning required is the removal of any suckers from around the base, as this form is usually grafted.

ASPECT: Full sun, partial shade
FLOWERING: February – March
HARDINESS: −10 to −15°C (14 to 5°F)

2–3 m (6–10 ft)

1.5–2.5m (5–8 ft)

Cotoneaster lacteus

COTONEASTER LACTEUS

Late cotoneaster

ROSACEAE

The late cotoneaster is a large evergreen shrub with dull green, deeply veined, leathery leaves that have a cream-coloured, finely hairy underside. In late autumn, large clusters of red fruits appear along the arching stems and usually last long into the winter. A sunny spot is best for good berry production.

However, I have found that the berries are a favourite treat for many birds and once they have found them, the fruits can then disappear rather quickly. I personally don't mind, though, as I have generally had time to enjoy them beforehand and appreciate that they offer birds a nutritious meal to help them get through the lean winter months.

This shrub's dense evergreen growth also makes it ideal for growing as a hedge.

Minimal pruning is needed if this graceful plant has been given enough space to grow to maturity, which is when it looks its best. Where needed, I prune lightly after flowering but leave some flowerheads to continue to develop into berries.

ASPECT: Full sun, partial shade
FRUIT INTEREST: October – January
HARDINESS: –10 to –15°C (14 to 5°F)

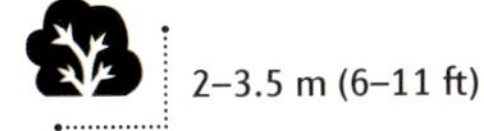

2–3.5 m (6–11 ft)

2–3.5 m (6–11 ft)

It would only be 2m if grown as a hedge

DAPHNE BHOLUA

Nepalese paper plant

THYMELAEACEAE

An evergreen, semi-evergreen or deciduous shrub, this daphne has an upright, open habit and has given rise to many cultivars. All produce clusters of highly fragrant four-lobed waxy flowers in colours from white to deep pink. They will do best in a sheltered spot and are relatively slow growing, so there is no need for pruning.

'Alba' produces pure white flowers that are intensely fragrant.

'Darjeeling' has very pale pink, almost white flowers, and is semi-evergreen, but retains some of its leaves all year, except in harsh winters. '

'Gurkha' is a deciduous form, which was bred from the variety *glacialis*, collected in the mountains of Nepal in the 1960s, and produces white flowers that open from rich pink buds.

'Jacqueline Postill' is the most popular cultivar and very floriferous, with highly fragrant clusters of purplish-pink flowers. This cultivar originated from a seedling of 'Gurkha' raised at the Hillier nursery in England.

ASPECT: Full sun, partial shade
FLOWERING: December – February
HARDINESS: –5 to –10°C (23 to 14°F)

Daphne bholua **'Gurkha'**

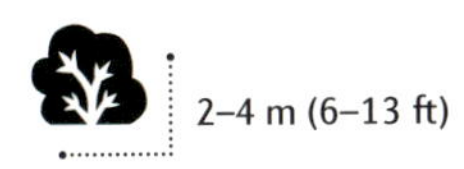

2–4 m (6–13 ft)

1.5–2.5 m (5–8 ft)

Daphne bholua **'Jacqueline Postill'**

DAPHNE LAUREOLA SUBSP. *PHILIPPI*

Dwarf spurge laurel

THYMELAEACEAE

This dwarf evergreen shrub has a small, rounded shape and produces glossy green leaves and clusters of greenish-yellow fragrant flowers between the leaf axils. It may not be as showy as some of the paler-flowered daphnes, but it will tolerate more shade than most and its compact size makes it ideal for smaller gardens, where the honey-scented perfume can be enjoyed.

No pruning should be required.

ASPECT: Full sun, partial shade

FLOWERING: February– March

HARDINESS: –10 to –15°C (14 to 5°F)

50 cm (20 in)

50 cm (20 in)

DAPHNE ODORA 'AUREOMARGINATA'

Golden-edged winter daphne

THYMELAEACEAE

This variegated evergreen form of *Daphne odora* is reasonably hardy and fairly compact. The glossy, green leaves have golden-yellow margins, giving it a striking appearance and year-round appeal. Fragrant pink winter flowers open from deep pink buds that are borne in small terminal clusters.

The similar cultivar Daphne odora **'Rogbret'** has the same pink flowers but wider creamy-yellow leaf margins.

No pruning should be required.

ASPECT: Full sun, partial shade

FLOWERING: January – March

HARDINESS: –5 to –10°C (23 to 14°F)

1–1.5 m (3–5 ft)

1–1.5 m (3–5 ft)

Daphne odora 'Rogbret'

Distylium racemosum

DISTYLIUM RACEMOSUM

Isu tree

HAMAMELIDACEAE

Not a common shrub, but this evergreen member of the witch hazel family is well worth growing if you want something a bit different. The plant blooms for a relatively short period from February to late March, and the unusual flowers lack true petals, being made up of red calyces (petal-like structures) and purple stamens. Each individual bloom is small but they are produced in clusters that stand out well against the dark green foliage.

Relatively slow growing, it will eventually make a medium-sized to large shrub. In its native Japan it grows into a large tree – and is best grown in partial shade since it is naturally a woodland plant.

The only pruning that is required is to remove any damaged or dead branches as you see them.

ASPECT: Partial shade

FLOWERING: February – April

HARDINESS: –5 to –10°C (23 to 14°F)

2–3 m (6–10 ft)

2.5–3 m (8–10 ft)

EDGEWORTHIA CHRYSANTHA

Paperbush

THYMELAEACEAE

Highly scented and related to the daphnes, this deciduous shrub has long been cultivated in its native Japan and China for its bark fibres, which are used to make fine handmade paper. Horticulturally, this attractive shrub is grown for its rounded clusters of bright yellow tubular flowers, which open from white silky-haired buds. The blooms appear on the tips of the bare stems in late winter and fill the air with their sweet fragrance.

'Grandiflora' is generally bigger in stature, with larger clusters of yellow flowers.

'Red Dragon' is a relatively new cultivar, which I am sure will become as popular as the yellow-flowered paperbushes, as the striking orange–red flowers of this plant are equally fragrant.

No pruning should be required.

ASPECT: Full sun, partial shade
FLOWERING: February – April
HARDINESS: –5 to –10°C (23 to 14°F)

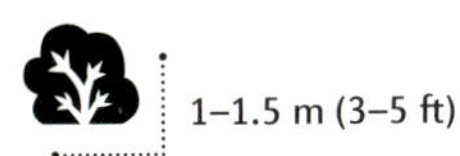
1–1.5 m (3–5 ft)

1–1.5 m (3–5 ft)

ELAEAGNUS PUNGENS 'MACULATA'
Oleaster 'Maculata'
ELAEAGNACEAE

One of the best shrubs for year-round attractive foliage, the leaves of this large, hardy evergreen have an irregular dark green margin with a bright yellow central splash of colour, each one displaying a slightly different pattern.

It is also a good choice for a coastal garden, where it will tolerate the salty air, and is often used as hedging and screening in maritime regions.

Carry out any pruning that is required in late spring, and always wear gloves since the stems can be slightly spiny. This shrub has a tendency to revert back to type and produce plain green foliage, so keep an eye out for this growth and remove it promptly, as the stems are more vigorous and can eventually take over.

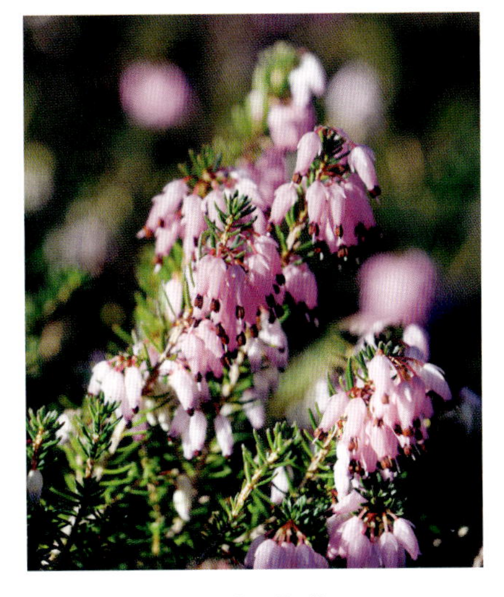

Erica carnea **'December Red'**

ASPECT: Full sun, partial shade
FOLIAGE INTEREST: Year round
HARDINESS: –10 to –15°C (14 to 5°F)

2.5–4 m (8–13 ft)

2.5–4 m (8–13 ft)

ERICA CARNEA
Winter heath
ERICACEAE

A low-growing evergreen shrub with a spreading habit, winter heath makes a good ground-cover plant, carpeting areas with its needle-like dark green leaves and bell-shaped flowers in winter. The blooms come in a variety of colours from pure white to deep pink. This little plant also looks good with winter-flowering bulbs and perennials in a container.

Winter heath will tolerate slightly alkaline conditions, but acid soil suits it better.

Cutting the stems back to just below the flowers once they have faded will ensure good new growth in the next flowering season.

'December Red' starts to flower in early winter, opening pink at first and deepening to a light purple as the flowers age. Very floriferous, it blooms over a long period. Height to 15 cm/6 in.

'Snowbelle' has large pure white flowers that spread along the stems, almost covering the foliage. Height to 20 cm/8 in.

'Winter Rubin' has dark foliage and a prostrate habit, with deep purple–red flowers covering the stems. Height to 15 cm/6 in.

ASPECT: Full sun, partial shade
FLOWERING: November – March
HARDINESS: –15 to –20°C (5 to –4°F)

10–50 cm (4–20 in)

10–50 cm (4–20 in)

Edgeworthia chrysantha

Elaeagnus pungens **'Maculata'**

ERICA X DARLEYENSIS

Darley Dale heath

ERICACEAE

This cross between *Erica carnea* and *Erica erigena* produces larger plants than its parents and forms small bush-shaped clumps. The beauty of this Erica is that it doesn't need an acid soil like most heathers and it is lime-tolerant, so will grow well in neutral soils. It tends to flower a little later than *Erica carnea*, but there are cultivars that bloom in midwinter too.

'Furzey' has a low, spreading habit and tolerates partial shade, making it an ideal heather for growing under and around deciduous trees and shrubs. The flowers are lilac pink. Height to 30 cm/12 in.

'Phoebe' produces clusters of rose-pink flowers along the stems that stand out well against the dark green foliage. It sometimes comes into flower as early as October. Height to 40 cm/16 in.

'Tweety' is grown more for its foliage than for the dark pink flowers, which are only produced sparingly. However, its yellow foliage is quite striking and takes on orange tints in winter. Height to 30 cm/12 in.

'White Perfection' is a strong-growing form and regarded as one of the best whites of this hybrid. Its pure white flowers are produced from early winter, while the new foliage is tipped with yellow in spring. Height to 45 cm/18 in.

Cut plants back to just below the flowers once they have faded to ensure good new growth for the next flowering season.

ASPECT: Full sun, partial shade

FLOWERING: December – April

HARDINESS: –10 to –15°C (14 to 5°F)

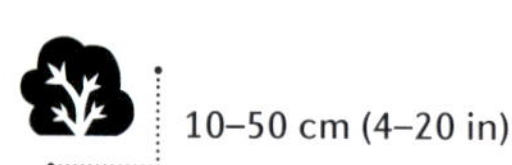

The unusual golden foliage of this heather is an added winter feature

Erica x darleyensis 'Tweety'

Erica x darleyensis 'Furzey'

Erica x darleyensis 'White Perfection'

Euonymus fortunei **'Emerald Gaiety'**

Euonymus fortunei **'Emerald 'n' Gold'**

EUONYMUS FORTUNEI

Winter creeper

CELASTRACEAE

This versatile shrub is evergreen, except in particularly cold winters or areas with generally harsh weather conditions, when it can lose its leaves (semi-evergreen). It is very adaptable and will grow in full sun or full shade, and can be used as a free-standing shrub, where it will root at intervals along its stems to cover the ground, or as a low hedge, since it tolerates regular clipping. The stems will also climb, so another option is to grow it on a wall or fence, or through other woody plants.

These *Euonymus* cultivars will generally make low-spreading mounds, but they will be taller when grown on or through structures, reaching heights of more than 2 m/6 ft.

There are many cultivated forms, but those listed here have stood the test of time and look particularly good in winter.

'Emerald Gaiety' is a small bushy shrub with pale green stems and rounded leaves that are mainly green but with irregular creamy-white margins that are often tinged pink in winter. Height to 1.5 m/5 ft.

'Emerald 'n' Gold' has a spreading habit and the leaves have green centres and yellow margins that are also often tinged pink in winter. Height to 1.5 m/5 ft.

'Silver Queen' is similar to 'Emerald Gaiety' but has larger dark green variegated leaves that are edged with wider expanses of creamy-white. Height to 2.5 m/8 ft.

Any pruning that is needed can be carried out from mid- to late spring.

ASPECT: Full sun, partial shade, full shade

FOLIAGE INTEREST: Year round – margins tinged pink in winter

HARDINESS: –10 to –15°C (14 to 5°F)

1–2.5 m (3–8 ft)

1–2.5 m (3–8 ft)

Euonymus fortunei **'Silver Queen'**

X *FATSHEDERA LIZEI*

Tree ivy

ARALIACEAE

The x hybrid symbol before this plant's name indicates that it is an inter-generic hybrid, which means a cross between two different genera. In this case, it is between *Fatsia japonica* (Japanese aralia) and *Hedera hibernica* (Irish ivy).

The result is a medium-sized evergreen shrub with leathery, palmate, glossy green leaves that lend a tropical look to a planting scheme. It will make a spreading mound if left to its own devices or allow it to scramble up trellis or through other plants. It produces round heads of pale cream flowers in late autumn, which are generally sterile, but occasionally have been known to produce purple berries in early winter.

The variegated cultivar **'Annemieke'**, which is sometimes found under the cultivar name **'Aureomaculata'**, is to my mind a better plant. The lighter yellow blotches in the centre of the leaves add interest and will bring to life a dark and shady spot in the garden.

Best grown in a sheltered position as the foliage can be damaged by frost, the only pruning that this shrub will need is to remove damaged leaves in late spring or early summer, after all risk of frost has passed.

ASPECT: Full sun, partial shade
FOLIAGE INTEREST: Year round
FLOWERING: November – March
HARDINESS: 1 to –5°C (34 to 23°F)

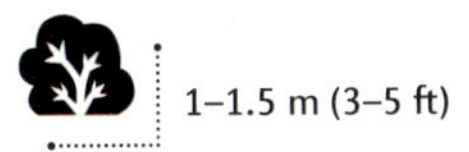

Fatshedera x *lizei*

FATSIA JAPONICA

Japanese aralia

ARALIACEAE

Although it looks very tropical, the Japanese aralia is, in fact, very hardy and a great addition to a winter garden, offering height and a different form to a shrub bed with its large, evergreen, glossy palmate leaves.

Flowering in late autumn, it produces spherical cream blooms followed in winter by attractive round clusters of black berries on long white branched stems. *Fatsia* is a great plant for the back of a border or use it to fill a corner. It is also a very good source of late-season nectar for many insects and the flowers are often almost entirely covered in pollinators when it is in bloom.

There is also a variegated form called **'Spider's Web'**, which has white mottling mainly around the leaf margins but sometimes spread across the whole surface.

Little pruning is required, other than to restrict the plant's size, as it can get quite large when growing in ideal conditions. Stems can be cut right down to the base, if needed, in late spring.

ASPECT: Full sun, partial shade
FOLIAGE INTEREST: Year round
HARDINESS: –5 to –10°C (23 to 14°F)

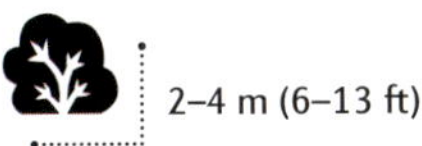

Fatsia japonica 'Spider's Web'

OPPOSITE
x *Fatshedera lizei* 'Annemieke'

Garrya elliptica 'James Roof'

GARRYA ELLIPTICA 'JAMES ROOF'

Silk tassel bush

GARRYACEAE

A large, bushy evergreen shrub with glossy green leaves that have wavy margins, the silk tassel bush is a must for any winter garden, when clusters of the amazing male flowers (known as silver tassels), which are around 20 cm/8 in in length, cover the plant. This spectacular shrub is a good choice for the back of a border, or as a stand-alone feature plant.

No pruning is necessary, unless it is needed to restrict its size, and this is best carried out after the catkins have faded.

ASPECT: Full sun, partial shade
FLOWERING: December – February
HARDINESS: –5 to –10°C (23 to 14°F)

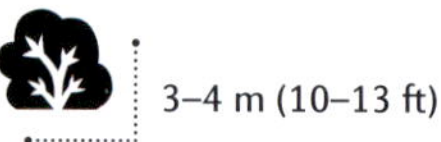
3–4 m (10–13 ft)

3–4 m (10–13 ft)

GAULTHERIA MUCRONATA

Prickly heath

ERICACEAE

Commonly found under the synonym name *Pernettya mucronata*, this hardy little evergreen shrub is native to Chile and Argentina and has a compact habit, with dense, dark green, glossy foliage with spiny tips.

It flowers in late spring, the small pinkish-white bell-shaped flowers eventually resulting in striking bright red shiny berries from late autumn, which remain showy all winter. It is only the female plants that produce berries, after the flowers have been pollinated.

Cultivars that produce pink and white berries are also available.

Its relatively small size makes it ideal for planting in containers as part of a winter display, and it goes particularly well with other acid-loving plants such as heathers, rhododendrons, and skimmias, although it will thrive in most soils.

Pruning in mid- to late spring is only needed for shaping, tidying, and removing any parts that have died.

ASPECT: Full sun, partial shade
FRUITING: December – March
HARDINESS: –10 to –15°C (14 to 5°F)

1–1.5 m (3–5 ft)

1–1.5 m (3–5 ft)

Gaultheria mucronata

Hamamelis

It is difficult to pick just one outstanding winter-flowering plant, but I think for most people, the witch hazel must be right up there in the top five.

There are four species of *Hamamelis*, two from Asia and two from North and Central America; one of the latter is *Hamamelis virginiana* (used for the witch hazel extract in cosmetic and pharmaceutical products), which flowers in the autumn. The other three all bloom in winter and early spring. There are also hybrids between the Japanese and Chinese witch hazels that have produced many of the most popular cultivars, which come in an amazing range of colours, forms, and scents. The spidery-looking flowers are unaffected by frost, too, and in some it enhances them as they glisten in the winter sun.

Their distinctive flowers with their fragrant, spidery-looking petals make them a popular winter shrub

Hamamelis mollis

The common name witch hazel doesn't have any reference to witches. In this case, it derives from 'wych', as in wych elm, an old English word meaning 'pliant' that refers to a shrub with flexible stems.

All *Hamamelis* are deciduous, but a few will retain some of their old leaves when in flower, but not every year.

The flowering twigs can be cut for a vase indoors, where they will provide a colourful, scented, long-lasting display.

HAMAMELIS JAPONICA – Japanese witch hazel – HAMAMELIDACEAE

The small clusters of mildly fragrant yellow flowers, with the distinctive twisted petals of all witch hazels, grow along the naked branches in the depths of winter. This species is also one of the parents, together with *Hamamelis mollis*, of the popular hybrids *Hamamelis* x *intermedia*.

HAMAMELIS MOLLIS – Chinese witch hazel – HAMAMELIDACEAE

One of the earliest witch hazels to come into flower, producing clusters of spicy-scented, yellow flowers in December, the species name *mollis* is Latin for soft, and refers to the felt-like feel of the leaves, which turn yellow in autumn. 'Brevipetala' is an upright form with yellow flowers, the petals of which have an orange–red base.

Hamamelis vernalis

HAMAMELIS VERNALIS – Ozark witch hazel – hamamelidaceae

The Ozark witch hazel produces the smallest flowers of all of these plants, but they appear in large quantities and are highly scented, which makes up for their diminutive size. The flower colours can be variable, but are generally coppery or pale orange. A few cultivars have been selected: 'Red Imp' with red and orange petals, and 'Squib', which has bright yellow petals.

Hammamelis x *intermedia* 'Angelly'

HAMAMELIS X *INTERMEDIA* – Hybrid witch hazel – HAMAMELIDACEAE

This species has the widest choice of varieties, with flower colours ranging from pale yellow to orange and red. Many also have a spicy or sweet fragrance.

'Angelly' has large primrose-yellow flowers and is very floriferous. Its spidery blooms almost completely cover the bare stems.

'Diane' is probably the best of the red forms. Its autumn leaf colour is also crimson–maroon.

'Jelena' has coppery-red flowers, is sweetly scented, and has the bonus of good autumn leaf colour.

As all of these cultivars are grafted, any suckers that are produced from below the graft union (knobbly section on the lower stem) should be removed to avoid reversion and weakening of the cultivar.

ASPECT: Full sun, partial shade
FLOWERING: December – March
HARDINESS: −10 to −15°C (14 to 5°F)

2–4 m (6–13 ft)

2–4 m (6–13 ft)

Hammamelis x *intermedia* 'Diane'

Hammamelis x *intermedia* 'Jelena'

Hydrangea febrifuga

HEPTACODIUM MICONIOIDES

Seven son flower tree

CAPRIFOLIACEAE

This is a large shrub that is extremely hardy but not commonly grown. It has an open, upright habit and can make a small tree if left unpruned for around 20 years. The fragrant white flowers appear in late autumn, but once the trunk and stems mature, the yellowish-brown bark flakes and peels in long strips, which I think is attractive during the winter after the leaves have fallen, especially when the shrub is underplanted with hellebores, winter aconites, and snowdrops.

Little pruning is necessary, except to remove damaged wood and possibly to restrict growth, but when choosing a site for this shrub, give it plenty of space to grow to maturity.

ASPECT: Full sun, partial shade
STEM INTEREST: Year round
HARDINESS: –10 to –15°C (14 to 5°F)

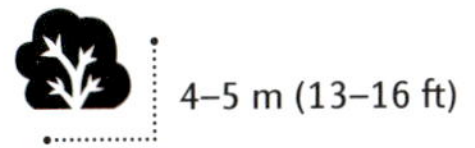

4–5 m (13–16 ft)

3–4 m (10–13 ft)

HYDRANGEA FEBRIFUGA

Chinese quinine

HYDRANGEACEAE

This small evergreen shrub was until recently found under the name *Dichroa febrifuga*, and may still be listed as that by nurseries, but it is still in the same family.

It is relatively rare in cultivation, but well worth growing for its winter interest. During the summer months it produces large heads of pink or blue starry flowers (occasionally both), followed in late autumn and throughout the winter by unusual dark blue fruits.

It is best grown in some shade and with protection from cold winter winds.

It is a low-maintenance plant, requiring minimal pruning, apart from the removal of dead or dying branches. Light pruning of other stems should be carried out in early spring.

ASPECT: Partial shade
FLOWERING: January – March
HARDINESS: 1 to –5°C (34 to 23°F)

0.5–1.5 m (20 in–5 ft)

0.5–1 m (20–39 in)

Heptacodium miconioides

JASMINUM NUDIFLORUM

Winter jasmine

CAPRIFOLIACEAE

Winter jasmine may appear to be evergreen because of its bright green stems but, in fact, it is deciduous. The small leaves are each made up of three leaflets (known as trifoliate), and fall at the end of autumn to reveal the long arching branches. Unlike many other jasmines, the winter jasmine does not have twining stems that help it to climb and support itself, and it will need to be tied to wires if grown as a wall shrub. However, left to its own devices, it will scramble over low walls, producing a cascade of flowers, or, when used as a stand-alone shrub, it will eventually make a sprawling mound, with the loose stems rooting where they touch the ground.

Although not fragrant, the masses of bright yellow flowers that are produced along its bare, shiny green stems more than compensate for the lack of scent. Its specific epithet *nudiflorum* means 'flowers coming before the leaves'.

Lonicera fragrantissima

***Lonicera* x *purpusii* 'Winter Beauty'**

It's pretty low maintenance, but any pruning that is necessary should be carried out in spring, after flowering.
ASPECT: Full sun, partial shade
FLOWERING: January – March
HARDINESS: –10 to –15°C (14 to 5°F)

2 –4 m (6–13 ft)

1– 2.5 m (3–8 ft)

LONICERA FRAGRANTISSIMA

Winter-flowering honeysuckle

CAPRIFOLIACEAE

This wonderful deciduous (occasionally semi-evergreen) winter-flowering shrub has a powerful, sweet fragrance that is not only attractive to us, but also to bees, particularly bumblebees venturing out on warmer, sunny winter days to sample its nectar.

The fragrant creamy-white flowers are usually borne in pairs along its stems just as the leaves are beginning to appear. It makes a bushy, free-standing, multi-stemmed shrub or you can train it against a wall.

You may find this plant listed as *Lonicera standishii*, which is a synonym for *L. fragrantissima*.

Pruning should be carried out in late spring after flowering. Remove older and weak wood right back to the base to encourage new flowering stems to form.
ASPECT: Full sun, partial shade
FLOWERING: December – March
HARDINESS: –10 to –15°C (14 to 5°F)

1.5–2 m (5–6 ft)

1.5–2 m (5–6 ft)

LONICERA X *PURPUSII* 'WINTER BEAUTY'

Winter honeysuckle

CAPRIFOLIACEAE

There is very little difference between this winter-flowering honeysuckle and *L. fragrantissima*, which is one of the parents of this hybrid cultivar. Both are deciduous or semi-evergreen shrubs with clusters of fragrant white flowers, but 'Winter Beauty' has darker purple flowering stems, and I think it is more floriferous, but flowering does differ a little from year to year.

This shrub looks best when the arching stems are left unpruned, but if any pruning is needed, cut it back after flowering in early spring.
ASPECT: Full sun, partial shade
FLOWERING: December – March
HARDINESS: –10 to –15°C (14 to 5°F)

1.5–2 m (5–6 ft)

1.5–2 m (5–6 ft)

OPPOSITE
Winter jasmine, *Jasminum nudiflorum*

MAHONIA X MEDIA

Oregon grape hybrid

BERBERIDACEAE

This cross between *Mahonia oiwakensis* subsp. *lomariifolia* and *Mahonia japonica* is an evergreen hybrid, which has produced some notable cultivars that make great architectural features in a winter garden. Planted at the back of a large border, the spiky, glossy green foliage is a good foil for smaller, lighter-coloured plants, while the upright spikes of small bell-shaped lemon-yellow flowers that rise above the foliage add to the plant's allure. Like many winter-flowering shrubs, this Mahonia is an invaluable food source for foraging insects, particularly bees. The grape-like purple berries that follow the flowers are also enjoyed by birds in late winter and early spring.

'Charity' has upright spikes of highly fragrant yellow flowers from late autumn through to late winter. Height: 2–4 m/ 6–13 ft.

'Winter Sun' is similar to 'Charity', but generally a bit smaller, so a better choice if space is limited. Height: 2–3 m/6–10 ft.

Pruning is generally only needed if plants become leggy, and this can be carried out from mid- to late spring, shortening stems back to side shoots lower down.

ASPECT: Partial shade, full shade
FLOWERING: November – February
HARDINESS: –10 to –15°C (14 to 5°F)

2–4 m (6–13 ft)

2 –3 m (6–10 ft)

***Mahonia* x *media* 'Winter Sun'**

NANDINA DOMESTICA

Heavenly bamboo

BERBERIDACEAE

As its common name suggests, this compact shrub has a similar habit and, to some extent, leaf shape to bamboo, but it actually belongs to the same family as *Berberis* and *Mahonia*.

While the evergreen foliage offers year-round interest, the plant looks its best in winter. The leaves change colour as they mature, red-flushed in spring as new foliage emerges; they then turn green in summer and finally take on burgundy and purple tints as temperatures fall in the autumn and winter. Tiny white flowers are produced in sprays in midsummer, which transform into bright red berries in autumn and remain on the plant through winter, alongside the fiery seasonal foliage.

There is also a yellow-berried form, which is unfortunately not widely available.

Any pruning should be carried out in spring; cut old and ragged stems right down to the base, which will encourage new stems to form.

ASPECT: Full sun
FRUITS: October – February
HARDINESS: 10 to 15°C (14 to 5°F)

Nandina domestica

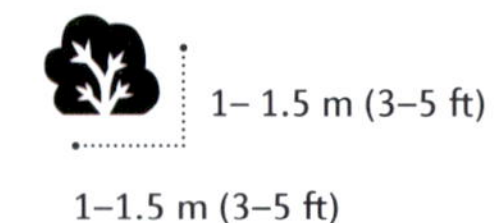

1– 1.5 m (3–5 ft)

1–1.5 m (3–5 ft)

PITTOSPORUM TENUIFOLIUM 'VARIEGATUM'

Tawhiwhi

PITTOSPORACEAE

A small, rounded, evergreen shrub with attractive variegated foliage, the leaves of this bushy plant are glaucous green with creamy-white wavy margins, often tinged with pale purple or pink during the winter, which contrasts well with the dark purple stems.

Like most evergreen shrubs it is very useful in a winter garden, injecting colour into a mixed planting and making an attractive screen or backdrop to other plants. Because of its small size, this *Pittosporum* is a good choice for areas where space is limited and will thrive in a container. It can also be used for topiary or grown as a low hedge.

Pruning is only needed to shape the plant, or to remove dead or diseased branches.

'Tom Thumb' is a dark purple form that works particularly well as a backdrop for showing off paler plants, but also as a specimen. Slow growing, it reaches a height of around 1 m/3 ft.

ASPECT: Full sun, partial shade

FOLIAGE INTEREST: Year round

HARDINESS: –10 to –15°C (14 to 5°F)

2–3 m (6–10 ft)

1.5–2.5 m (5–8 ft)

PYRACANTHA

Firethorn

ROSACEAE

Most of the firethorns best suited to winter gardens are the wide range of cultivars, all of which are evergreen, very hardy and tolerant of shade, making them ideal for growing in a north- or east-facing position. They also make great wall-trained shrubs, but are equally attractive when grown as specimen plants. Their sharp, spiny stems create effective security hedges and screens, too.

Pittosporum tenuifolium **'Variegatum'**

Ideal for a smaller garden or space. This purple form of Pittosporum is compact and slow growing

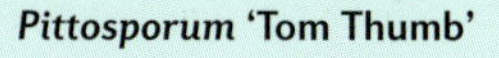

Pittosporum **'Tom Thumb'**

In late spring they produce masses of pretty white flowers, but it is the fruit that offers winter interest. Depending on the species or cultivar, the brightly coloured berries appear in late autumn and can last well into winter, except in particularly harsh weather, when many different bird species will feast on them. The fruits are not usually the birds' first choice but occasionally they will strip some of them from the plant. However, the berries appear in such profusion that there are usually plenty for both the birds and the gardener to enjoy.

Personally, I like to grow firethorns as wall shrubs, either pruned tightly against a vertical structure and tied in, or free-standing against a wall and lightly pruned.

Always wear robust gloves when pruning a *Pyrancantha* to protect your hands and arms against the long, sharp, and very tough thorns. Cut them back after flowering, shortening the non-flowering shoots to ensure berries for the winter.

Fireblight can be an issue. As its name suggests, branches affected by this

Pyracantha **'Saphyr Orange'**

bacterial disease look like their leaves have been scorched by fire, while the foliage on other branches may look perfectly normal. Prune out growth showing any signs of the disease, cutting stems back to at least 30 cm/12 in beyond the infection to healthy wood. There are resistant cultivars, including the 'Saphyr' range.

Orange-, red- and yellow-berried forms are available, and here are a few of the best.

'Orange Glow' is probably the best-known cultivar of all of the firethorns and has been around for decades. It is large

Pyracantha rogersiana **'Flava'**

and very spiny, with dark green glossy leaves; the flowers appear in late spring and early summer, followed by masses of long-lasting bright orange berries.

'Soleil d'Or' has thorny reddish stems and lighter green foliage than many other cultivars. It produces huge clusters of bright yellow berries in autumn, which are usually left by winter-feeding birds until after the red and orange varieties are finished.

***Pyracantha koidzumii* 'Victory'** is one of the best firethorns for red berries. It has a low, spreading habit but also does very well when trained as a wall shrub. Both its clusters of small white late spring flowers and the bright red berries stand out well against the glossy green leaves.

***P. rogersiana* 'Flava'** is another yellow-berried variety but with smaller, narrower leaves than many of the others, and the fruits seem somewhat glossier.

ASPECT: Full sun, partial shade
FRUITS: October – February
HARDINESS: –10 to –15°C (14 to 5°F)

1.5–3 m (5–10 ft)

1.5–3 m (5–10 ft)

Pyracantha **'Orange Glow'**

OPPOSITE
Pyracantha **'Soleil d'Or'**

RHODODENDRON MUCRONULATUM
Korean rhododendron

ERICACEAE

The solitary funnel-shaped flowers of the Korean rhododendron appear from winter to early spring on the bare stems of this medium-sized, deciduous shrub and really stand out when grown in shady areas of the garden.

The flower colour is quite variable, with blooms ranging from lilac to white. The most popular is a selection named as **'Cornell Pink'**, which has clear pink flowers. It can become a bit straggly but pruning after flowering will produce a stronger, bushier plant.

Light pruning is all that is required for both species and cultivars, removing any crossing and congested branches, or those that are dead or damaged. Cut plants back in late spring or early summer.

ASPECT: Partial shade
FLOWERING: January – March
HARDINESS: –10 to –15°C (14 to 5°F)

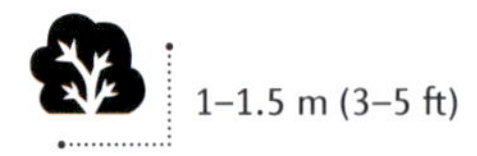

1–1.5 m (3–5 ft)

1–1.5 m (3–5 ft)

RHODODENDRON 'PRESIDENT ROOSEVELT'

ERICACEAE

The first time I saw this plant, the unusual foliage left me unsure of what it was, and I certainly didn't suspect a rhododendron.

Each of the dark green glossy leaves has a central splash of bright creamy-yellow, which creates an eye-catching feature – you definitely won't walk past this shrub without a second look.

While the dramatic foliage is the key feature in a winter garden, a bonus in early spring is the trusses of bright red flowers with pinkish-white centres that accompany it.

Like most rhododendrons it is relatively slow growing, and because it is also small in stature, little if any pruning is needed. Cut it back after flowering in late spring, if required.

ASPECT: Partial shade
FOLIAGE INTEREST: Year round
FLOWERING: March – April
HARDINESS: –10 to –15°C (14 to 5°F)

1.5–2.5 m (5–8 ft)

1.5–2.5 m (5–8 ft)

RIBES LAURIFOLIUM
Laurel-leaved currant

GROSSULARIACEAE

I was drawn to this wonderful small evergreen shrub from the first time I saw it, and will always have a space in my garden for it.

The low, arching chestnut-coloured stems and dull green leathery leaves are in winter accompanied by tubular yellowish-green flowers, borne in pendant clusters, which can differ in size, depending on whether the plant is male or female. The female plants tend to have longer, but sparser flowers and will occasionally produce red fruits that ripen to black after flowering.

Rhododendron mucronulatum

***Rhododendron* 'President Roosevelt'**

Ribes laurifolium

Rubus cockburnianus

Pruning to tidy and remove any dead material should be carried out after flowering in late spring.

ASPECT: Full sun, partial shade
FLOWERING: January – March
HARDINESS: –10 to –15°C (14 to 5°F)

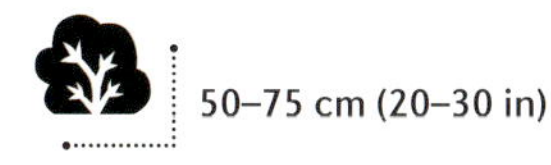

50–75 cm (20–30 in)

1–1.5 m (3–5 ft)

RUBUS COCKBURNIANUS

White-stemmed bramble

ROSACEAE

Grown for its stand-out winter stem colour, *Rubus cockburnianus* makes a striking feature plant.

Its tall, arching, thorny stems are covered with a white bloom and look spectacular throughout the winter months, especially when planted in small groups among other shrubs, perennials, and bulbs that flower at this time. To guarantee bright stems each winter, cut them all down to the base in spring, which will encourage the plant to produce new white canes while also restricting its size.

***Rubus* 'Goldenvale'**

ASPECT: Full sun, partial shade
STEM INTEREST: January – March
HARDINESS: –10 to –15°C (14 to 5°F)

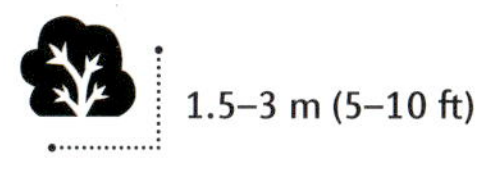

1.5–3 m (5–10 ft)

1.5– 3 m (5–10 ft)

RUBUS COCKBURNIANUS 'GOLDENVALE'

Golden-leaved bramble

ROSACEAE

This cultivar of *Rubus cockburnianus* has similar white bloom-covered winter stems to the species, but it is not quite as tall, and has a more spreading habit. You can also see the pinkish stems beneath the white bloom, but the main attraction of this cultivar is the golden-yellow fern-like young foliage (hence its cultivar name) that follows on in early spring.

Remove the older stems as new ones are produced to encourage young growth with bright stem colour the following winter.

ASPECT: Full sun, partial shade
STEM INTEREST: December – March
HARDINESS: –10 to –15°C (14 to 5°F)

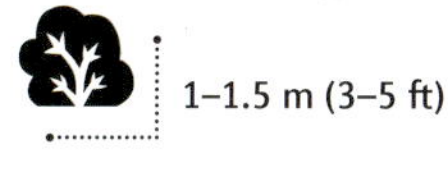

1–1.5 m (3–5 ft)

1–1.5 m (3–5 ft)

RUBUS THIBETANUS

Ghost bramble

ROSACEAE

The ghost bramble is also known as the silver fern because of its whitish-silver, fern-like, pinnate leaves, but it is also grown for its beautiful winter stem interest.

After losing its foliage in late autumn, this deciduous shrub's long arching stems, which have been purplish all summer, become covered in a white ghostly bloom, hence the common name. The stems are also covered with small prickles and form a dense thicket.

The plant's pale stems look great when grown alongside dogwoods and willows in contrasting colours that set each other off.

The ghost bramble will also do well planted on a steep bank or as a defensive screen, since it needs little maintenance, apart from the removal of old or dead stems to keep it healthy and encourage new young growth. Cut it back in early spring just before the new leaves appear when you can see all the stems clearly.

Rubus thibetanus

ASPECT: Full sun, partial shade
STEM INTEREST: December – March
HARDINESS: –10 to –15°C (14 to 5°F)

1–1.5 m (3–5 ft)

1–2 m (3–6 ft)

SARCOCOCCA CONFUSA

Sweet box

BUXACEAE

Also known as the Christmas box, as it usually flowers from late December, this medium-sized evergreen shrub has dark green glossy leaves and creamy-white flowers with protruding stamens. The winter blooms are sweetly scented and borne in small clusters in the leaf axils. They are followed by shiny round black berries, which often appear at the same time as the flowers. This is a really useful plant for a gloomy spot as it is extremely tolerant of deep shade.

The natural habit can be spoilt by pruning, so only remove damaged material as and when you see it.

ASPECT: Partial shade, full shade
FLOWERING: December – March
HARDINESS: –10 to –15°C (14 to 5°F)

1 –1.5 m (3–5 ft)

1–1.5 m (3–5 ft)

Sarcococca confusa

SARCOCOCCA HOOKERIANA VAR. *DIGYNA*

Sweet box

BUXACEAE

Like *Sarcococca confusa*, this evergreen shrub also produces small clusters of fragrant creamy-white flowers, followed by black berries, but its narrow leaves are longer and held on purplish stems. It tolerates deep shade, too, and is ideal for difficult areas that receive little or no sun throughout the day

'Winter Gem' is a hybrid form (*S. hookeriana* var. *digyna* 'Purple Stem' x *S. hookeriana* var. *humilis*) with large glossy green leaves and highly scented creamy-white flowers that open from pink buds and retain this colour at the base of the flowers. It will slowly spread by suckering and is shorter than the species. Height: 60 cm/24 in.

The natural habit can be spoilt by pruning, so only remove damaged material as and when you see it.

ASPECT: Partial shade, full shade
FRUITS: October–March
HARDINESS: –10 to –15°C (14 to 5°F)

1–1.5 m (3–5 ft)

1–1.5 m (3–5 ft)

Sarcococca hookeriana* var. *digyna

***Skimmia* x *confusa* 'Kew Green'**

***Skimmia japonica* 'Rubella'**

SKIMMIA X *CONFUSA* 'KEW GREEN'

RUTACEAE

A tough little evergreen, this compact shrub will tolerate a shady corner in the garden that receives a few hours of sun each day. It produces tight green buds in midwinter. In late winter these open into terminal clusters of creamy-white fragrant flowers in late winter that persist well into spring, but as it is a male clone it will not produce berries after flowering.

Its glossy green foliage makes it an attractive plant even when not in flower, and it will eventually grow into a neat, mounded shape.

Any necessary pruning should be carried out in spring.

ASPECT: Partial shade
FLOWERING BUDS: November – March
HARDINESS: –10 to –15°C (14 to 5°F)

0.5–1 m (20–39 in)

1–1.5 m (3– 5 ft)

SKIMMIA JAPONICA 'RUBELLA'

RUTACEAE

This small evergreen shrub is grown for its dark red flower buds, which develop in the autumn and persist right through the winter, contrasting well with the dark green foliage. They eventually open in early spring to reveal small white fragrant flowers. Its small compact size makes it ideal for growing in a container on its own or with other winter-flowering plants.

This skimmia is a male clone so will not produce berries after flowering.

Any necessary pruning should be carried out in spring.

ASPECT: Full sun, partial shade
FLOWERING BUDS: November – March
HARDINESS: –10 to –15°C (14 to 5°F)

0.5–1 m (20–39 in)

0.5–1 m (20–39 in)

SKIMMIA REEVESIANA

Japanese skimmia

RUTACEAE

Many skimmias are dioecious, which means the male and female flowers are on separate plants and both sexes are needed for the females to develop berries each year, but *Skimmia reevesiana* is hermaphrodite and self-fertile, reliably bearing fruits without a pollination partner.

This compact evergreen has leathery, dark green, glossy leaves, fragrant white flowers in spring and bright red berries from late autumn that last until the following spring.

Its size makes it ideal for a small garden, or for growing in a container.

Any necessary pruning should be carried out in spring.

ASPECT: Partial shade, full shade

FRUITS: October – March

HARDINESS: –10 to –15°C (14 to 5°F)

0.5–1 m (20–39 in)

0.5–1 m (20–39 in)

Stachyurus praecox

SOPHORA MICROPHYLLA 'SUN KING'

Kowhai

FABACEAE

An unusual-looking bushy evergreen shrub, in winter it produces large clusters

of bright golden-yellow pea-like flowers with yellow anthers. The glossy green leaves are made up of 30–40 pairs of small oblong leaflets and contrast well with the blooms.

The species is native to New Zealand (it is the country's unofficial national tree), where the Māori name *kōwhai* comes from and refers to the colour yellow. In New Zealand it flowers from late summer to late autumn but in Britain it blooms over a long period from late winter into early spring.

Kowhai is relatively hardy, but best grown with some protection from cold winds; it makes a beautiful wall shrub in a sheltered, sunny position.

Very little, if any, pruning is necessary; just remove dead or damaged branches in spring after flowering.

ASPECT: Full sun

FLOWERING: February – March

HARDINESS: –5 to –10°C (23 to 14°F)

1.5–2.5 m (5–8 ft)

1.5–2.5 m (5–8 ft)

STACHYURUS PRAECOX

Early stachyurus

STACHYURACEAE

A striking large deciduous shrub with a spreading habit, the early stachyurus is grown for its showy flowers which appear from late winter into early spring. From a distance the pendent racemes of small, pale yellow, bell-shaped flowers look like catkins, but on closer inspection, they are more like strings of pearls, hanging from the arching purple–brown branches.

This hardy, versatile shrub will make a statement in a bed or a border underplanted with bulbs and groundcover, or try training it against a wall or fence, where it will also thrive.

When grown as a free-standing shrub,

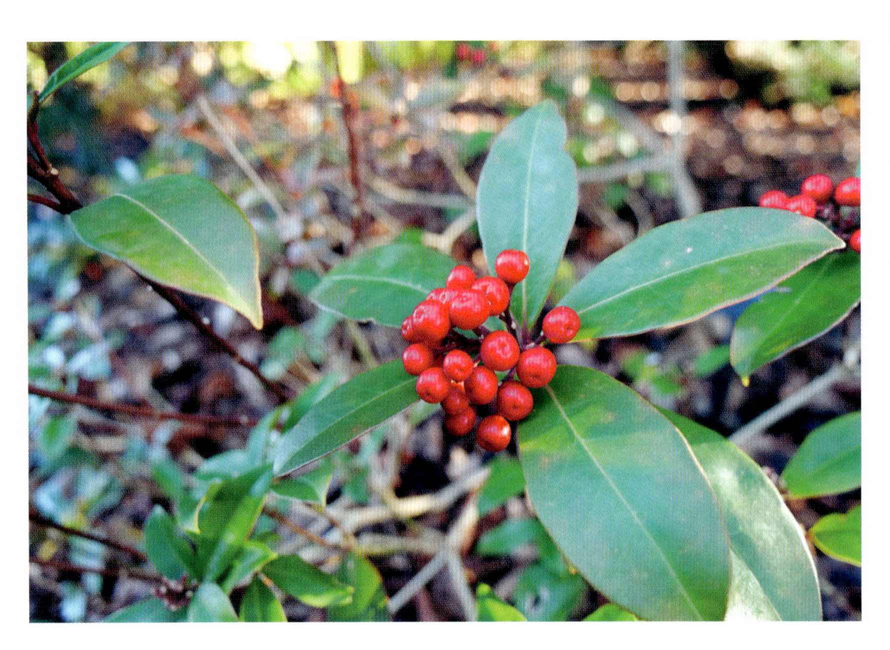

Skimmia reevesiana

minimal pruning is needed, so that it maintains its natural shape. If grown as a wall shrub, tie in strong stems and prune after flowering. Young stems that shoot up from the base should be used to replace some of the older growth from time to time to keep the plant flowering well in both free-standing and wall shrubs.

ASPECT: Partial shade
FLOWERING: February – April
HARDINESS: –10 to –15°C (14 to 5°F)

1.5–3 m (5–10 ft)

2–3 m (6–10 ft)

SYCOPSIS SINENSIS

Chinese sycopsis

HAMAMELIDACEAE

This large evergreen or semi-evergreen shrub, often referred to as a small tree, is rarely grown in gardens, probably because it is a large plant and only produces its unusual flowers for a short period in late winter. But if space isn't an issue, then it is definitely worth growing.

The unusual petal-less flowers are borne in dense clusters along the branches, each bloom emerging from within a reddish-brown cupped bract (petal-like structure) and consisting of prominent red anthers on long yellow filaments.

It needs little, if any, pruning, apart from the removal of dead or diseased branches in early spring after flowering.

ASPECT: Full sun, partial shade
FLOWERING: February – March
HARDINESS: –5 to –10°C (23 to 14°F)

3–5 m (10–16 ft)

3–4 m (10–13 ft)

Sycopsis sinensis

SYMPHORICARPOS ALBUS

Snowberry

CAPRIFOLIACEAE

Introduced and naturalised over much of Europe, the snowberry is actually native to the US.

It is a deciduous suckering shrub that can quickly create a dense thicket, making it ideal for screening or a wildlife hedge – the nectar-rich spring flowers are very attractive to bees and butterflies, while the snow-white berries are eaten by birds during the winter. *Symphoricarpos* is also known as the pearl tree, which I like because once the leaves have fallen, the pearl-like berries hang on long thin stems like clusters of stringed pearls.

Its suckering habit can make it quite difficult to control, but pruning out dead stems after flowering is all that is required.

Care should be taken as the berries are poisonous to humans.

ASPECT: Full sun, partial shade
FRUITS: October – March
HARDINESS: –10 to –15°C (14 to 5°F)

1.5–2 m (5–6 ft)

1.5–2 m (5–6 ft)

Symphoricarpos albus

VIBURNUM X BODNANTENSE

VIBURNACEAE

This hybrid cross between *Viburnum farreri* and *Viburnum grandiflorum* is a medium-sized to large deciduous shrub with upright stems, dark green leaves – bronze when young – and the scented flowers of both its parent plants. It has also led to the selection of some award-winning cultivars that mainly differ in their flower colour.

'Charles Lamont' has pale pink flowers, similar to those of *Viburnum farreri*.

'Dawn' is probably the most famous selection; it has clusters of white flowers tinged pink, with darker anthers, opening from red buds.

Viburnum x bodnantense

Viburnum x *bodnantense* 'Charles Lamont'

Viburnum x *bodnantense* 'Deben'

'Deben' has the whitest flowers of these three cultivars, opening from pink buds.

Prune out older stems from the base to make way for younger, more productive growth to maintain a good show of flowers each year.

ASPECT: Full sun, partial shade
FLOWERING: November – March
HARDINESS: –10 to –15°C (14 to 5°F

3 m (10 ft)

2 m (6 ft)

VIBURNUM FARRERI

Farrer viburnum

VIBURNACEAE

A favourite in any winter garden, this tall upright deciduous shrub produces clusters of flowers over a long period, opening in late autumn and continuing to bloom throughout the winter and early spring. The sweetly fragrant tubular pink

Viburnum farreri 'Candidissimum'

Viburnum tinus

flowers appear intermittently along the woody stems and fill the air with a rich almond scent. In spring, they accompany the new foliage, which has a coppery tone when it first unfurls.

'Candidissimum' is a distinctive form, with rounded clusters of pure white fragrant flowers, which grows taller than the species to around 3 m/10 ft.

Prune out older stems from the base to make way for younger flowering shoots, since the old wood weakens with age and does not flower well.

ASPECT: Full sun, partial shade
FLOWERING: November – March
HARDINESS: –10 to –15°C (14 to 5°F)

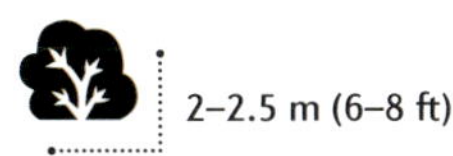

2–2.5 m (6–8 ft)

2–2.5 m (6–8 ft)

VIBURNUM NUDUM

Withe rod

VIBURNACEAE

This deciduous shrub has an upright habit and forms a rounded bush, roughly as tall

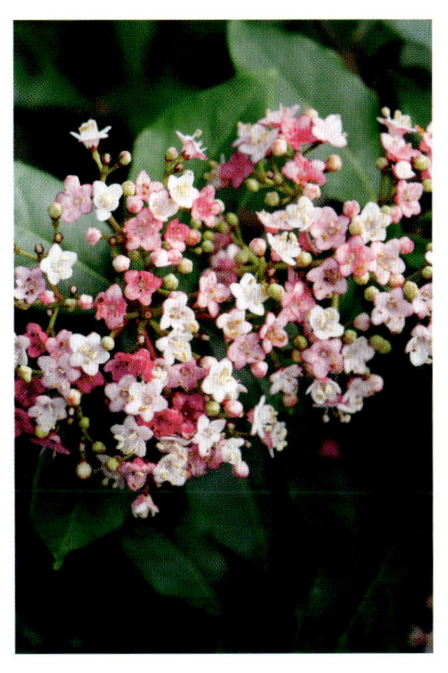

Viburnum tinus **'Eve Price'**

as it is wide, with glossy green leaves that turn crimson in autumn before they fall. The creamy-white clusters of flowers that appear in midsummer go on to produce berries from late autumn that gradually change colour from green to pink and blue as they mature, finally becoming almost black.

The cultivars make the best garden plants. **'Pink Beauty'** is an excellent self-fertile form that produces large clusters of berries that change colour from pale green to white, pink, and eventually to black, with several colours often appearing on the plant at the same time.

Little pruning is necessary, but any that is required should be carried out from mid- to late spring.

ASPECT: Full sun, partial shade
FRUITS: November – February
HARDINESS: –10 to –15°C (14 to 5°F)

2–2.5 m (6–8 ft)

2–2.5 m (6–8 ft)

VIBURNUM TINUS

Laurustinus

VIBURNACEAE

So many of the winter-flowering shrubs that we grow are from parts of Asia, particularly China and Japan, but this viburnum is from the Mediterranean and not only is it hardy in our climate, it also tolerates a wide variety of aspects, from full sun to full shade, as well as summer drought.

It makes a large evergreen shrub, with stems covered in dark green leaves that can be pruned or clipped regularly to make a great flowering hedge. The flowerheads are made up of many individual small creamy-white blooms that open from pink-tinged buds over a long period and are followed by attractive blue–black iridescent berries.

'Eve Price' is a more compact, denser plant, with pale pink flowers that open from dark pink buds. Height to 2 m/6 ft.

'Lisarose' has buds that are a deep brick-red and open to reveal white flowers tinged with pink. Height to 2.5 m/8 ft.

These viburnums respond well to hard pruning in spring after flowering, but this is only needed if plants outgrow their space.

ASPECT: Full sun, partial shade, full shade
FLOWERING: January – April
HARDINESS: –5 to –10°C (23 to 14°F)

3 m (10 ft)

3 m (10 ft)

Trees

ACACIA DEALBATA
Blue wattle
FABACEAE

Commonly called wattle, *Acacia* species naturally grow in warm tropical and sub-tropical parts of the world, with the majority native to Australia. However, they are surprisingly tough and hardy, and able to cope with several degrees of frost.

The blue wattle makes a small tree or large evergreen shrub, with grey–green feathery foliage and, from late winter to early spring, attractive fluffy pompoms of bright yellow fragrant flowers. It will flower most profusely in an open, sunny site.

The common name wattle refers to its use by early European settlers in Australia as wattle and daub for buildings.

Light pruning can be carried out after flowering in late spring.

ASPECT: Full sun, partial shade
FLOWERING: February – April
HARDINESS: 1 to −5°C (34 to 23°F)

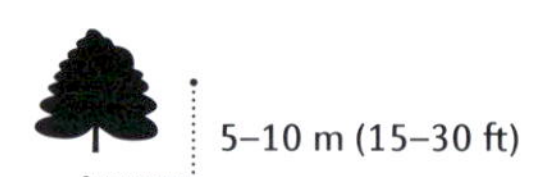

5–10 m (15–30 ft)

4–6 m (13–20 ft)

Acacia dealbata

Acer davidii

ACER DAVIDII
Snake-bark maple
SAPINDACEAE

Acer davidii makes a beautiful medium-sized deciduous tree and is one of the best of the so-called snake-bark maples. Offering year-round interest, the small lime-green to yellow flowers appear in spring and are followed by winged fruits in autumn when its leaves also turn orange and yellow. But in a winter garden it is the stunning bark of the trunk and stems that is its main feature.

The young stems are red (retaining this colour in winter), and then turn olive–green, often still with some red tones, as they mature, with eye-catching white vertical stripes.

Growing it as a multi-stemmed tree will restrict its height.

ASPECT: Full sun, partial shade
STEM INTEREST: Year round
HARDINESS: −10 to −15°C (14 to 5°F)

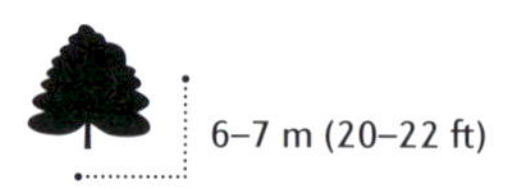

6–7 m (20–22 ft)

4–5 m (13–16 ft)

ACER GRISEUM
Paperbark maple
SAPINDACEAE

This small deciduous tree has the most beautiful cinnamon-coloured peeling bark that lends year-round interest to the garden, while its fiery autumn leaf colour is the most colourful of all the maples grown for their stem interest. Although the peeling papery bark is attractive throughout the seasons, it is in winter that the trunk and stems can really be appreciated. As it peels, it curls into thin rolls that remain attached but reveal a lighter tan-coloured bark beneath, producing an attractive multi-coloured affect.

Relatively slow growing, a multi-stemmed specimen will be showier, especially if space is limited.

ASPECT: Full sun, partial shade
STEM INTEREST: Year round
HARDINESS: −10 to −15°C (14 to 5°F)

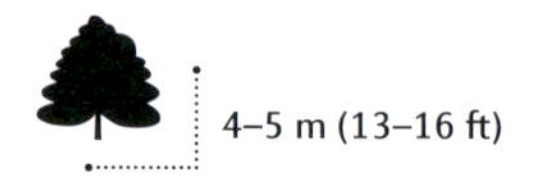

4–5 m (13–16 ft)

3–4 m (10–13 ft)

ARBUTUS X ANDRACHNOIDES
Hybrid strawberry tree
ERICACEAE

Arbutus x *andrachnoides* is a medium-sized evergreen tree, grown for its cinnamon-brown flaking bark, which is the main attraction in winter. However, it has merits at other times of year, too,

Arbutus* x *andrachnoides

most notably the bell-shaped white flowers that are produced during the autumn or spring, and occasionally through the winter.

The hybrid strawberry tree is a cross between the Grecian strawberry tree (*Arbutus andrachne*), and the strawberry tree (*Arbutus unedo*), and it arises naturally in the wild wherever the two parent species occur together.

It gets its attractive flaking bark from its Grecian parentage, and the colour becomes more vivid when the stems are wet after rain.

ASPECT: Full sun
STEM INTEREST: Year round
HARDINESS: –5 to –10°C (23 to 14°F)

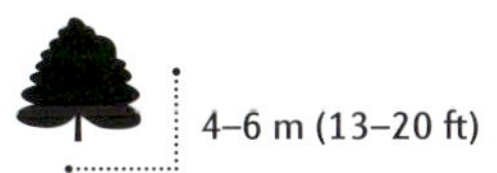

4–6 m (13–20 ft)

3–5 m (10–16 ft)

Acer griseum

AZARA MICROPHYLLA

Box-leaf azara

SALICACEAE

This evergreen eventually makes a small tree but is often classified as a large shrub. Its small rounded dark green leaves resemble those of box, hence the common name, but the clusters of small yellow flowers, almost insignificant singly but standing out en masse against the dark foliage, create its distinctive look. The blooms are highly scented, filling the air over a wide area with their incredible perfume and stopping you in your tracks. Different people smell different things. Most often it is said to be vanilla-scented, but I, like many, often smell chocolate. It is definitely a scent that will be noticed and makes a talking point.

'Variegata' has glossy dark green leaves with cream-coloured margins.

ASPECT: Full sun, partial shade
FLOWERING: February – April
HARDINESS: –5 to –10°C (23 to 14°F)

Azara microphylla

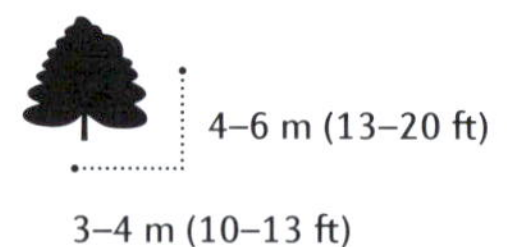

***Betula utilis* subsp. *albosinensis* 'Fascination'**

BETULA UTILIS SUBSP. *ALBOSINENSIS* 'FASCINATION'

Chinese silver birch

BETULACEAE

The Chinese silver birch, or red-barked birch, is a medium-sized deciduous tree that is known for its striking winter stem colour, which includes the creamy-white older bark that peels off in horizontal strips to reveal the pink, orange, and cinnamon-coloured new bark beneath. It also produces beautiful pale yellow 15 cm/6 in-long catkins in early spring, or late winter if the weather is mild, which are particularly attractive in the spring sunshine.

Snowdrops and winter aconites combine well with this birch, and many plants will also grow in the dappled shade beneath the boughs later in the year when these trees are in full leaf.

ASPECT: Full sun, partial shade
STEM INTEREST: Year round
HARDINESS: –15 to –20°C (5 to –4°F)

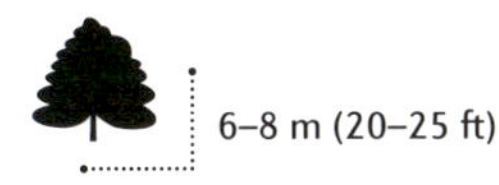
6–8 m (20–25 ft)
3–5 m (10–16 ft)

BETULA UTILIS SUBSP. *JACQUEMONTII* 'DOORENBOS'

Himalayan birch 'Doorenbos'

BETULACEAE

This medium-sized tree has one of the whitest barks of all of the birch clones. Each year the old layers peel off the trunk and branches to reveal gleaming white new bark and darker horizontal lenticels (pores). In autumn the leaves also turn golden-yellow, making it a tree for all seasons, not just for winter. Planted in a group of three or more with a contrasting background and underplanted with dogwoods (*Cornus*) or grasses, the bark

***Betula utilis* subsp. *jacquemontii* 'Doorenbos'**

colour will really stand out, creating a spectacular display.

ASPECT: Full sun, partial shade
STEM INTEREST: Year round
HARDINESS: –15 to –20°C (5 to -4°F)

6–8 m (20–25 ft)

3–5 m (10–16 ft)

CORNUS MAS

Cornelian cherry

CORNACEAE

Eventually making a small deciduous tree, the relatively slow-growing Cornelian cherry will reach 4–5 m/13–16 ft in time. In late winter and early spring its naked branches appear to turn yellow, which from a distance make it really stand out. On closer inspection, you will see that the colour is actually produced by small clusters of tiny bright lemon-yellow flowers with yellow flower stalks and protruding white stamens.

Adding to its charms are the red, cherry-like fruits (hence its common name) that follow the flowers in summer. They are edible but quite tart, so are best used in jams, jellies, and preserves. It also has very good autumn colour, the leaves taking on shades of yellow and reddish-purple.

A multi-stemmed specimen is a good option if space is limited and will grow to around half the height of a single-stemmed tree. Prune it during the dormant season in winter if it gets too large.

ASPECT: Full sun, partial shade
FLOWERING: February – April
HARDINESS: –15 to –20°C (5 to –4°F)

2.5–5 m (8–16 ft)

2.5–4 m (8–13 ft)

Cornus mas

Fragrant white flowers on long stems can have up to 20 petals, with flowering times varying from late winter to early summer

Drimys winteri

CRATAEGUS MONOGYNA 'BIFLORA'

Glastonbury thorn

ROSACEAE

The first time I became aware of this tree was in 2010 when one growing close to Glastonbury Tor had been vandalised and I was asked to propagate it by grafting from the remains of its damaged branches. Legend has it that the original tree grew from the hawthorn staff thrust into the ground by Saint Joseph of Arimathea on Christmas day, and it is also known as the holy thorn.

While this may be a myth, what is remarkable about this small, thorny, deciduous tree is that it occasionally flowers twice (as its cultivar name suggests). It blooms in winter and then produces another heavier flush of creamy-white flowers in April and May, which is when most common hawthorns come into flower.

ASPECT: Full sun, partial shade
FLOWERING: February – April
HARDINESS: –15 to –20°C (5 to -4°F)

3–5 m (10–16 ft)

2–3 m (6–10 ft)

DRIMYS WINTERI

Winter's bark

WINTERACEAE

This unusual evergreen small tree or large shrub is said in most literature to flower from late spring to early summer, but, in the south of England, it reliably blooms every year from January.

It is relatively hardy but is best given some protection by growing it against a south- or west-facing wall.

Its large glossy dark green leaves are oblong in shape and glaucous on the underside, while the loose terminal clusters of creamy-white, fragrant flowers, which open from pink buds, stand out well against the foliage.

Winter's bark is named after Captain John Wynter, who accompanied Sir Francis Drake on his round-the-world voyage. He discovered that the bark of this South American tree was rich in vitamin C and offered a cure for scurvy,

which was killing many of the crew. The bark was dipped in honey to mask its acrid taste.

ASPECT: Full sun, partial shade
FLOWERING: January – April
HARDINESS: –5 to –10°C (23 to 14°F)

5–10 m (16–30 ft)

3–6 m (10–20 ft)

EUCALYPTUS PAUCIFLORA SUBSP. *NIPHOPHILA*

Snow gum

MYRTACEAE

Most eucalyptus have colourful bark, the tones heightened when wet, but I have chosen this sub-species of snow gum because of its size. Most trees in this genus grow into huge trees that are too big for small or medium-sized gardens, whereas this one is much smaller, growing to just 5 m/16 ft. An ornamental evergreen, it has attractive flaking bark in different shades, from silvery–green to grey and cream, and, like all eucalyptus, the grey–green foliage is scented.

Any pruning should be carried out in early spring. This tree can be cut back hard to around 30 cm/12 in from the ground; coppicing it like this will produce a bushy plant with attractive foliage, but it will also remove the attractive flaking bark that comes with maturity.

ASPECT: Full sun, partial shade
STEM INTEREST: Year round
HARDINESS: –10 to –15°C (14 to 5°F)

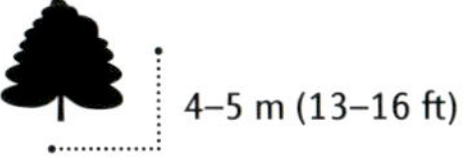

4–5 m (13–16 ft)

3–4 m (10–13 ft)

Eucalyptus pauciflora* subsp. *niphophila

ILEX

Hollies

There are so many different species, clones, and cultivars of holly that there have been whole books written about them. Most are evergreen, although some are deciduous, and the majority have foliage with spiny margins, while the prickles on a few, such as *Ilex aquifolium* 'Ferox Argentea', cover the whole leaf surface. Others, including cultivars of *Ilex* x *altaclerensis*, have smooth leaf margins and no spines at all. There is also a range of hollies with variegated leaves in a broad spectrum of colours, and a choice of berries, from the palest yellow to almost black.

Hollies are dioecious, which means that they produce male and female flowers on separate plants, and only the females bear the berries, while the male plants are needed for pollination to ensure that the fruits develop.

These small trees offer year-round interest, particularly the variegated evergreen types (the variegation is always better when plants are grown in full sun), but the main attraction in autumn and winter is the colourful berries. However, they don't always last long, since they are equally attractive to many different types of bird, which feed on the bounty while stocking up for the winter. In the UK, flocks of winter migrant birds such as fieldfares and redwings feast on them, particularly the bright red fruits, which are always the first to disappear; the orange and yellow berries often remain on the trees until spring.

And with so many to choose from, there is a holly to suit every garden.

Some variegated hollies tend to revert back to their natural green foliage, so keep an eye out for green-leaved.

Ilex decidua

Variegated holly, showing the all-green reversion, which needs pruning out when seen

Deciduous hollies

These hollies are known as winterberries and most are quite shrubby, but some will make small trees. Their berries are mainly bright red, but some cultivars have golden, orange, or pale yellow fruits. Like their evergreen counterparts, deciduous hollies have separate male and female plants and you must choose a female if you want berries.

There are many different species of deciduous holly, but the three below are the best for growing in most garden situations. Surprisingly, though, they are still not common in cultivation and you will need to search for them, but they make a great feature in a winter garden and are worth the effort.

The showy berries of the holly are a popular choice with florists

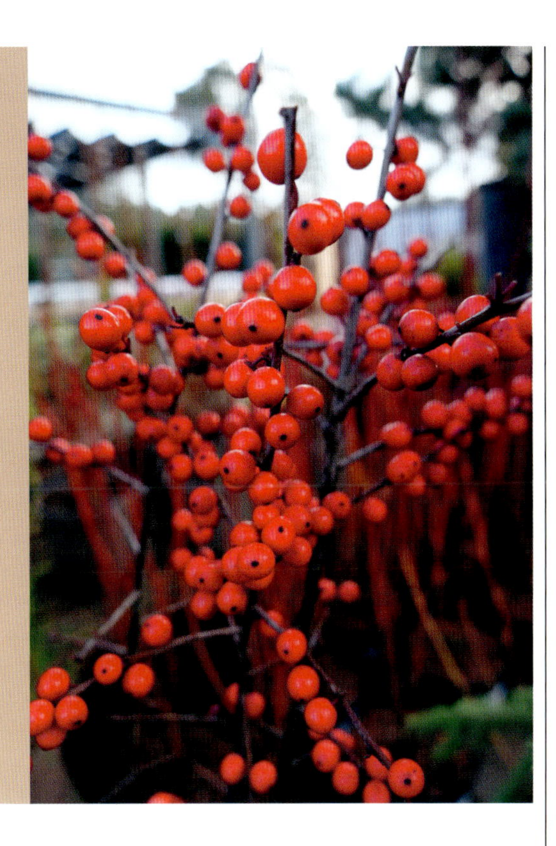

Ilex verticillata

Ilex decidua, commonly known as the possumhaw in its native America, is the most tolerant of these three deciduous hollies and able to cope with a wide range of conditions. It grows naturally in the south-eastern States in damp areas beside streams and rivers, but it is able to survive in quite dry soils too. It is usually sold as a multi-stemmed shrub but can be grown as a single-stemmed small tree to around 5–6 m/16–20 ft.

The fruits are most commonly dark red, but can be orangey-red. There are also cultivars with yellow and golden fruits, but these can be hard to track down. The berries form in the autumn before the leaves fall but persist long into the winter.

Ilex serrata, or Japanese winterberry, is from Asia, as the name suggests, occurring in China and Japan. It's generally smaller than the other deciduous hollies, reaching not much more than 3 m/10 ft in height. Not as hardy as its North American cousins – it should be hardy down to around –5°C (23°F) – it is therefore best grown in a sheltered, shady spot, where it will get some protection both from full sun and heavy frost, which can damage the fruits. In the spring it produces white or pale pink flowers, occasionally purple, and in autumn small clusters of red berries, which are showy even when competing with the foliage, but come into their own once the plant is completely leafless.

Ilex verticillata is another holly from North America and the most widely available deciduous form. It produces the most prolific fruits of all the winterberries and includes the cultivars 'Winter Red' and 'Winter Orange', which are named after the colours of their berries. They are all 1.5 to 2.5 m/ 5 to 8 ft tall.

Clusters of small greenish-white flowers appear in early summer, followed in autumn by bright berries, just as the foliage turns yellow, contrasting well with the fruits. This species tends to sucker so take this into consideration when choosing a planting position.

Retaining its berries after the twigs are cut, this holly makes a colourful addition to winter flower arrangements.

ASPECT: Full sun, partial shade
BERRIES: October – March
HARDINESS: –10 to –15°C (14 to 5°F)

3–6 m (10–20 ft)

3–4 m (10–13 ft)

Evergreen hollies

ILEX X ALTACLERENSIS
– Highclere holly – AQUIFOLIACEAE

This group of hollies is the product of a cross between the common holly, *Ilex aquifolium*, and *Ilex perado*, the Madeira holly. Commonly known as the Highclere holly, this hybrid was first described from a plant growing on the Highclere estate near Newbury, Berkshire, in south-east England.

The hybrid and its cultivars have larger leaves, with fewer and smaller spines along their margins, than their common holly parent, while many are spineless like those of the Madeira holly.

They are vigorous trees, many too large for the average-sized garden, with the tallest reaching around 20 m/70 ft in height, but most will grow to less than half this size and can be useful as a tall screening hedge.

They are particularly resistant to pollution and a good choice for planting in coastal areas, since they are very tolerant of salt-laden air.

'Golden King' is one of the best variegated and most decorative cultivars, with glossy oval dark green leaves that have beautiful spineless golden-yellow margins. The young stems are flushed with purple and produce small white flowers in early summer, followed by tight clusters of reddish-brown berries in autumn and winter.

The almost spineless leaves and variegated margins of this holly have made it a popular garden choice

'Golden King'

This is a good holly for brightening up a dark corner, and grows to just 5 m/16 ft in 20 years.

Height to 6 m/20 ft.

'James G. Esson' produces an abundance of bright red berries in large clusters along pale purplish stems that contrast well with the glossy dark green leaves, which are almost spineless, most having smooth margins.

Height: 8 m/26 ft.

'Lawsoniana' is similar to 'Golden King' but the variegation is reversed, with the yellow colouring in the centre of the leaves, although they are very varied, the yellow variegation sometimes covering most of the leaf. The dark red berries contrast well with the brightly coloured leaves.

Height: 6 m/20 ft.

'W. J. Bean' is named after a 20th-century curator of the Royal Botanic Gardens, Kew. This slow-growing cultivar stays compact, and is an ideal choice for the smaller garden. The large leaves have deeply wavy margins that are spiny, unlike many of the Highclere hybrids. The berries are bright red and a good size. This cultivar may be difficult to find, but it is well worth hunting down.

Height to 3 m/10 ft.

ASPECT: Full sun, partial shade
BERRIES: October – March
HARDINESS: –10 to –15°C (14 to 5°F)

4–10 m (13–30 ft)

3–4 m (10–13 ft)

'James G. Esson'

'W. J. Bean'

ILEX AQUIFOLIUM

Common holly

AQUIFOLIACEAE

As its name suggests, this is the UK's commonest holly and the most widely grown in British gardens. It is also native to the British Isles, as well as many other parts of Europe, and is found growing naturally in many of our woodlands as an understorey tree and, as such, is perfect for the shadier areas of a garden.

Cultivars and varieties of the common holly (of which there are dozens) usually make small trees, depending on which are chosen, but, beware, as some can reach up to 20 m/70 ft in height. In a garden setting, this species can be clipped into topiary, trimmed as a hedge, or grown as a stand-alone specimen tree.

As with all hollies, female plants produce the berries, which follow the fragrant nectar-rich flowers in May and June that are attractive to many insects, particularly bees. The berries range in colour from yellow, orange, and most commonly bright red.

This tree has long been associated with Christmas, and has been used as a natural decoration for wreaths and adorned Christmas cards for over a century.

'Bacciflava' is the best yellow-berried cultivar and produces glossy green leaves that are variably spiny or sometimes almost spineless. This cultivar has been around for at least 300 years and is also known as 'Fructu Luteo'. The berry colour means that they are usually left alone by birds until all the red-berried forms are exhausted. Height to 6 m/20 ft.

'Ferox Argentea' is a male holly and, as such, doesn't bear any berries in the autumn and winter, but it is a useful cultivar for pollinating nearby female plants. Even without the berries, the spiny variegated foliage with creamy-white margins and purple twigs give this small evergreen tree year-round appeal. The spines are produced on the leaf surface as well as the margins, giving rise to this plant's common name, the hedgehog holly. Height to 4 m/13 ft.

***Ilex aquifolium* 'Ferox Argentea'**

'J. C. van Tol' has the bonus of being a self-fertile female holly, producing berries without a nearby male. As with most of the *I. aquifolium* cultivars, the young stems are purple, and it has an upright habit, with dark green glossy leaves that are almost spineless. It bears an abundance of bright red fruits in autumn and winter. Height to 6 m/20 ft.

'Silver Milkboy' has variegated leaves with creamy-white or pale yellow central markings. The green leaf margins are wavy and spiny. Both male and female forms are available, but the male is more popular, grown for its distinctive foliage. The berries are at best sparse on the female forms. Height to 6 m/20 ft.

ASPECT: Full sun, partial shade
BERRIES: October – March
HARDINESS: −10 to −15°C (14 to 5°F)

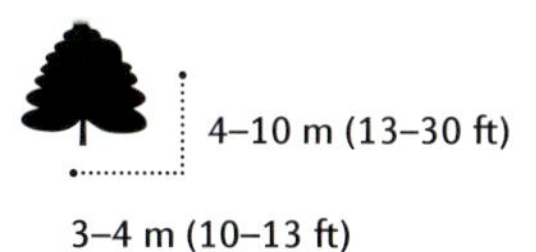

4–10 m (13–30 ft)
3–4 m (10–13 ft)

PARROTIA PERSICA

Persian ironwood

HAMAMELIDACEAE

This small deciduous tree or large shrub has a spreading habit and is noted for its attractive peeling bark and amazing autumn colour, when the leaves turn fiery shades of yellow, orange, red, and purple. The foliage also often hangs on into early winter, until just after the first frosts or snowfall.

Later in winter its bare twigs are clothed in small clusters of deep red, spider-like, petal-less flowers with attractive red stamens that draw attention to this beautiful tree.

ASPECT: Full sun, partial shade
FLOWERS: January – March
HARDINESS: −10 to −15°C (14 to 5°F)

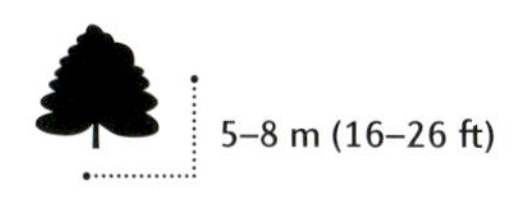

5–8 m (16–26 ft)
4–8 m (13–26 ft)

Parrotia persica

***Parrotia* flowers**

Pinus bungeana

PINUS BUNGEANA
Lacebark pine

PINACEAE

An unusual, attractive pine, this tree is still fairly rare in cultivation, probably because it is very slow-growing and takes up to ten years before the attractive peeling bark starts to have any real impact. However, once mature, it is a real stunner.

The lacebark pine eventually forms a medium-sized tree that produces branches from low down on the main trunk, creating a network of beautiful stems with peeling bark as they age. The older bark is silver–grey and white with red and purple tones, and flakes off in small plates in a similar way to plane trees (*Platanus* x *hispanica*), to reveal the brownish young growth beneath.

ASPECT: Full sun, partial shade
STEM INTEREST: Year round
HARDINESS: –10 to –15°C (14 to 5°F)

5–8 m (16–26 ft)

4–8 m (13–16 ft)

PINUS CONTORTA 'CHIEF JOSEPH'
Lodgepole pine 'Chief Joseph'

PINACEAE

This dwarf evergreen conifer is noted for its stunning winter foliage. Through the spring and summer it is yellowish green, but as temperatures begin to drop it turns an eye-catching golden-yellow, standing out in the drab winter landscape. The needle-like foliage also has a slight twist, adding to its attractiveness.

The cultivar is named after the famous leader of a North American tribe who lived in the Wallowa Mountains in Oregon, which is where this form was discovered in the late 1970s.

It is very slow-growing, increasing in height up to 15 cm/6 in per year.

ASPECT: Full sun
FOLIAGE INTEREST: Year-round (from December to March)
HARDINESS: –10 to –15°C (14 to 5°F)

2–5 m (6–16 ft)

1.5–2.5 m (5–8 ft)

Pinus contorta 'Chief Joseph'

PINUS MUGO 'DEZEMBER GOLD'

Dwarf mountain pine

PINACEAE

Unlike many pines that grow to great heights, the dwarf species *Pinus mugo* rarely reaches more than 2 to 3 m/6 to 10 ft, and is a good choice for a small garden or a large container.

It is slow-growing, with an upright habit, and the cultivar 'Dezember Gold' produces slightly twisted golden-green needle-like leaves, with butter-yellow new growth in early spring.

'Winter Sun' is similar to 'Dezember Gold' but generally smaller overall, forming a rounded mound. During the cold winter months its foliage turns bright golden-yellow and because of its small size it is particularly suited to growing in containers or on a rockery. Height 50 cm – 1 m/20 – 39in.

ASPECT: Full sun

FOLIAGE INTEREST: Year round

HARDINESS: –10 to –15°C (14 to 5°F)

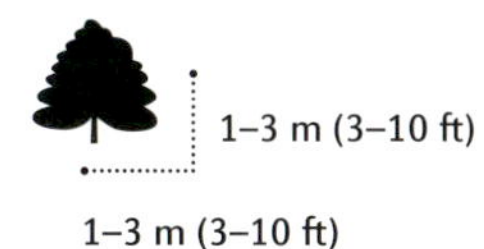

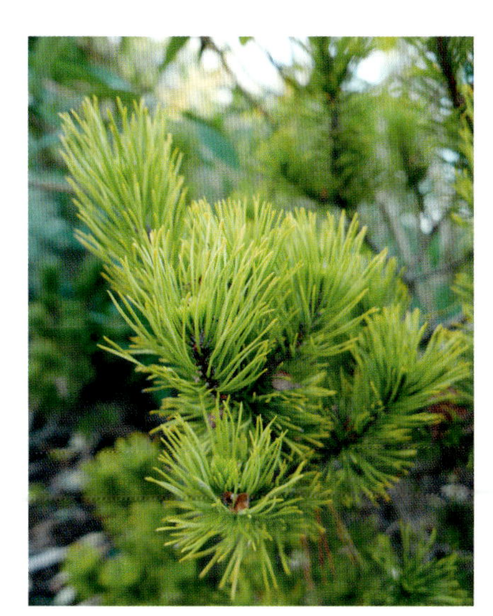

***Pinus mugo* 'Dezember Gold'**

PRUNUS CERASIFERA 'PISSARDII'

Cherry plum

ROSACEAE

Most references suggest that the flowering time of this small tree begins in early spring (March). This may have been the case a decade or so ago, but it now blooms reliably in February.

A deciduous, medium-sized tree with a rounded habit, its dark twiggy branches are almost black and help to show off the masses of five-petalled pale pink flowers, each with long showy anthers. As the flowers mature, they become pure white. The main flush of blooms appears just before the dark purple leaves unfurl, the colourful foliage giving rise to the tree's other common name, the purple-leaved plum.

The early flowers are also a welcome source of early nectar and pollen for bees.

ASPECT: Full sun

FLOWERING: February – April

HARDINESS: –10 to –15°C (14 to 5°F)

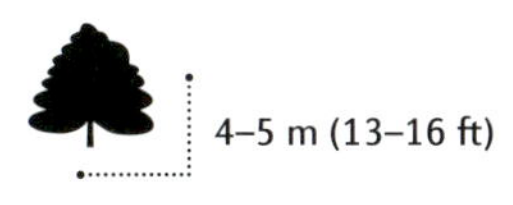

***Pinus mugo* 'Winter Sun'**

***Prunus cerasifera* 'Pissardii'**

PRUNUS INCISA 'PRAECOX'

Fuji cherry 'Praecox'

ROSACEAE

A small deciduous tree that flowers freely from late winter, this Fuji cherry bears two or four individual white flowers on each stalk. The blooms are pale pink in bud and some of this colour carries over into the flowers, giving them a pinkish-white appearance. Most cherries are noted for their autumn colour and this one is no exception, the green leaves transforming into fiery tones of red and orange before they fall. Grow this Fuji cherry in full sun for the best winter flowers and autumn colour.

ASPECT: Full sun, partial shade

FLOWERING: February – March

HARDINESS: –10 to –15°C (14 to 5°F)

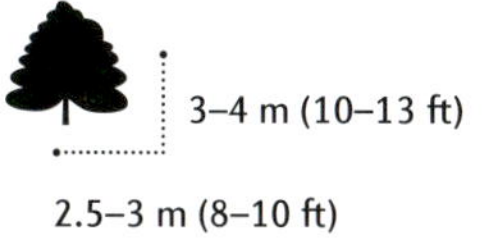

***Prunus incisa* 'Praecox'**

PRUNUS MUME 'BENI-CHIDORI'

Japanese apricot 'Beni-chidori'

ROSACEAE

This tree's colourful, scented winter flowers and compact habit make *Prunus mume* 'Beni-chidori' a showstopper for a small or medium-sized garden. The deep pink single flowers appear from late winter to early spring and the almond-scented blossom can be better appreciated if you plant this little Japanese apricot in a large container on a sunny patio, which will allow you to get up close to the blooms. Potted trees are also easier to move to a sheltered spot in winter to protect their early flowers from frost damage.

Pruning should be carried out in late spring or summer, if needed.

ASPECT: Full sun
FLOWERING: February – March
HARDINESS: –10 to –15°C (14 to 5°F)

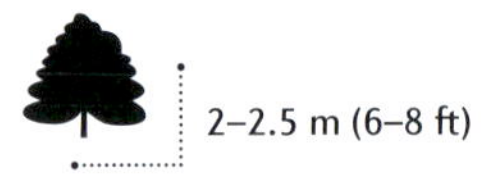

2–2.5 m (6–8 ft)

2–2.5 m (6–8 ft)

PRUNUS SERRULA

Tibetan cherry

ROSACEAE

Another good choice for a small to medium-sized garden, in most situations the relatively slow-growing Tibetan cherry will reach around 5 m/16 ft in height after 20 to 30 years when grown on a single stem, or remain shorter if grown as a multi-stemmed tree. It has rich, shiny, dark red mahogany-coloured bark that is often said to resemble polished mahogany. As the tree matures, the bark peels off in bands, which is why it is also known as the paperbark cherry.

The eye-catching bark gleams in the winter sun and looks particularly effective when grown next to the lighter-coloured stems of a yellow dogwood (*Cornus*).

Prunus serrula

It also has the bonus of good autumn colour when the leaves turn a lovely rich yellow.

ASPECT: Full sun, partial shade
STEM INTEREST: Year round
HARDINESS: –15 to –20°C (5 to –4°F)

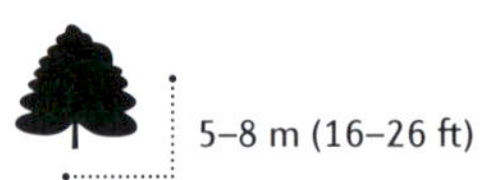

5–8 m (16–26 ft)

4–6 m (13–20 ft)

Prunus* x *subhirtella

PRUNUS X *SUBHIRTELLA*

Winter-flowering cherry

ROSACEAE

Also known as the spring cherry, since it has a relatively long flowering period from late autumn/early winter until early spring, the leafless stems of this deciduous tree are clothed in small clusters of beautiful rose-pink flowers that become paler as they age. The first of the flowers often appear as early as November, but the main flush occurs after Christmas, usually between February and March, and provides a welcome, cheerful sight to brighten the greyest days. As with many cherries, *Prunus* x *subhirtella* produces good autumn colour, when the foliage turns shades of yellow, orange, and red.

Small cherries are often produced and enjoyed by birds in early summer.

ASPECT: Full sun
FLOWERING: November – March
HARDINESS: –15 to –20°C (5 to –4°F)

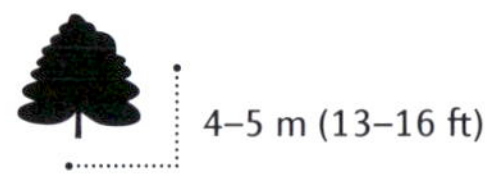

4–5 m (13–16 ft)

4–5 m (13–16 ft)

SALIX ALBA SUBSP. *VITELLINA* 'YELVERTON'

Golden willow

SALICACEAE

Salix alba, or the white willow, is a native of the UK and often seen pollarded and growing along the banks of streams and rivers. While this species has yellowish stems, the popular golden willow features brighter orange–red new growth.

Both need to be hard pruned in early spring every couple of years to promote the formation of young stems that provide the glowing winter stem colours. Cutting it back will also restrict the plant to a manageable size. If left unpruned, they will lose their vibrant colours, except at the tips of the branches, and become large trees.

The height given here is for a mature tree, but coppiced or pollarded plants can be kept to any desired height, from 1.5 m/5ft to their mature size.

ASPECT: Full sun

STEM INTEREST: January – March

HARDINESS: −10 to −15°C (14 to 5°F)

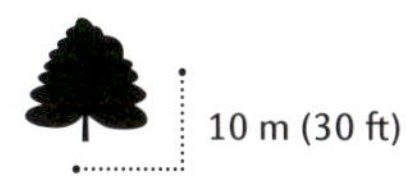

SALIX GRACILISTYLA 'MOUNT ASO'

Pink pussy willow

SALICACEAE

The fluffy deep pink catkins that appear on the bare stems of this willow are quite amazing and I remember the first time I saw it I had a double take and couldn't quite believe it was real. But real it is.

If left to its own devices, this small deciduous tree will grow to around 3 m/10 ft tall, but to get the best from its unusual colourful catkins it needs to be pollarded regularly, which will restrict its height so that they can easily be seen easily and enjoyed.

Pollarding produces a canopy of short, straight stems on a clear trunk. To achieve this, remove the lower branches until you have a clear trunk of the desired height. Then simply cut back hard all the flowering stems (those that have had catkins on) above this point, so the tree looks like an upright stem with a large knotty fist on top. Do this after flowering each or every other year to encourage healthy young stems and a good display of catkins.

ASPECT: Full sun

CATKINS: January – March

HARDINESS: −10 to −15°C (14 to 5°F)

***Salix alba* subsp. *vitellina* 'Yelverton'**

Pink pussy willow

***Tilia cordata* 'Winter Orange'**

TILIA CORDATA 'WINTER ORANGE'

Small-leaved lime 'Winter Orange'

MALVACEAE

For most of the year, the red buds and colourful orange–red twigs of this small-leaved lime are hidden beneath the green foliage, but once the leaves have fallen, striking one- and two-year-old shoots are revealed that really stand out on sunny winter days.

The best way to enjoy these colourful twigs is to pollard the tree regularly to promote lots of new colourful shoots (see *Salix gracilistyla* 'Mount Aso'). You can cut the tree down to at any height, but a clear stem of about 1.5 to 2m/5 to 6ft is generally best – prune back to around 50 cm/20 in below the final height of the clear stem you want to allow for new growth. This tree can also be stooled or coppiced in a similar way to dogwoods by cutting all the stems down to 5–10 cm/2–4 in from the ground, or to the previous year's stubs.

Cut the tree back every two years in early spring.

ASPECT: Full sun, partial shade

STEM INTEREST: December – March

HARDINESS: −10 to −15°C (14 to 5°F)

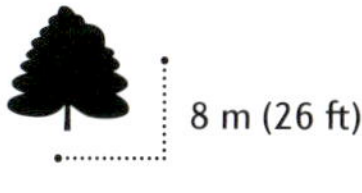

Glossary

Alternate Leaves arranged at each node on different sides of the stem. Not opposite
Anther The pollen-bearing male reproductive organ
Axil The junction of leaf and stem
Axillary Arising in the axil
Basal Located at the base of the plant or stem; usually refers to leaves
Biennial Germinates and grows in its first year; flowers and completes its life cycle in its second year
Bract A modified leaf, usually at the base of a flower. Sometimes more showy than the actual flower
Bulb An underground storage organ made up of swollen leaf bases
Calyx The outer parts of a flower, the sepals
Cambium The layer of cells below the bark of roots and stems that divides to produce new tissue
Climber Also known as lianas, these are plants with long trailing stems modified to climb by hooks/thorns, suckers, and tendrils
Corm Resembling a bulb, but replaced with new growth annually on top of the old one
Corona The cup-shaped floral structure between the tepals and the stamens; for example, the trumpet on a narcissus
Culm A stalk or stem of various grasses, sedges, and reeds
Cultivar A new plant produced in cultivation rather than in the wild, which has to be propagated vegetatively to remain true
Cuticle Waxy surface of a leaf, limiting water loss and protecting against extreme high and low temperatures
Cyathium A typical inflorescence in the euphorbia genus
Deciduous A plant that sheds its leaves at the end of its growing season, renewing them at the start of the next season
Dioecious Having male and female flowers on separate plants
Endemic Restricted in the wild to a specific geographic region
Epiphyte A plant that does not have roots in the ground, but grows on the surface of other plants
Evergreen A plant that retains most of its leaves throughout the year
Filament A hair-like structure that supports the anther, which together make up the stamen
Floriferous Plant producing many flowers
Genus A group of species with similar characteristics, e.g. Hamamelis
Glaucous Bluish-grey in colour
Globular Spherical or rounded in shape
Habitat Place where a plant grows, which is often characteristic of a species
Herbaceous A plant that is not woody and dies back to around ground level during the winter
Hybrid A plant resulting from cross-fertilisation of two different individual species
Inflorescence The flowering part of a plant
Lanceolate Having a narrow outline, usually widest in the middle, tapering towards the tip. Lance-shaped
Lateral At the side
Lax Loose and spreading
Linear Long, thin, grass-like leaves
Margin Outer edge, usually of a leaf
Monocarpic Flowering once and then dying
Monoecious Plants with male and female flowers on the same plant, but separate
Native A plant growing in a place where it was not introduced by humans or animals
Node The place on a stem where the leaves are attached
Obovate Oval, but widest above the mid-point and narrower at the base
Panicle A branching stem of stalked flowers
Perennial Any plant living for three years or more. Usually flowering annually, but not always in their first year
Petiole The stalk that joins a leaf to a stem
Photosynthesis The process by which plants use sunlight to convert carbon dioxide and water into the sugars they use as food
Pinnae A primary division of a pinnate leaf
Pinnate With leaflets arranged either side of single stalk
Prostrate Growing flat, or close to the ground
Reversion To change back to a previous form or colour

Rhizome A creeping stem, which can be above or below the ground
Rosette A cluster of leaves radiating from the same point, usually basal
Semi-evergreen Plants that lose their leaves for a short period of time between the old foliage falling and the new foliage growth
Sepal A segment of the calyx
Sessile Stalkless
Shrub A perennial with woody, many-branched stems
Species A class of plants whose members have similar characteristics to each other
Spine A rigid sharp structure on a stem or leaf
Stamen The male part of a flower, consisting of both the filament and anther
Stigma The top of the style, which receives the pollen
Style The slender stalk in a flower that connects the stigma and the ovary
Sub-shrub A small plant with a woody base and herbaceous top
Tendril A thread-like modified leaflet, used to anchor plants and help them to climb
Tepal Used to describe the outer petal-like parts of a flower when petals and sepals look alike. Usually in the flowers of bulbous plants, e.g. daffodils
Tree A perennial, usually with a single woody stem
Tuber A swollen underground stem or root
Umbel An inflorescence with several flowers all arising from the same point
Undulate With a wavy margin or surface
Variety Plants that often occur naturally and usually come true from seed. See also cultivar
Vegetatively Propagating a plant from parts other than its reproductive part, i.e. roots or stems
Whorl The arrangement of flowers or leaves that circle around the same point of the stem
Woody perennial See shrub and tree

Further reading

Bourne, Val (2006).
The Winter Garden,
CASSELL ILLUSTRATED, LONDON, ENGLAND.

Brownlee, George G. (2021).
A Passion for Snowdrops: a personal perspective.
WHITTLES PUBLISHING, DUNBEATH, SCOTLAND.

Cox, Freda (2019).
A Gardener's Guide to Snowdrops.
THE CROWOOD PRESS, MARLBOROUGH, ENGLAND.

Gregson, Sally (2022).
Designing and Creating a Winter Garden,
THE CROWOOD PRESS, MARLBOROUGH, ENGLAND.

Hardy, Emma (2019).
The Winter Garden.
CICO BOOKS, LONDON, ENGLAND.

Lancaster, Roy (2019).
The Hillier Manual of Trees & Shrubs.
RHS MEDIA, LONDON.

McAllister, Hugh & Marshall, Rosalyn (2017).
Hedera: The Complete guide.
RHS MEDIA, LONDON.

Websites

Gardeners' World Magazine: www.gardenersworld.com
Plants For A Future: pfaf.org
Royal Botanic Gardens, Kew, Plants of the World Online: powo.science.kew.org
Royal Horticultural Society: rhs.org.uk
Tree and Shrubs Online: treesandshrubsonline.org

Flowering by month

Prunus x subhirtella

DECEMBER	
Camellia sasanqua cultivars	Grasses
Chaenomeles x *superba* cultivars	*Hamamelis* spp. and cultivars
Chimonanthus praecox	*Hedera* spp.
Clematis cirrhosa	*Helleborus* 'Walberton's Rosemary'
Clematis napaulensis	*Lonicera fragrantissima*
Clematis 'Winter Beauty'	*Lonicera* x *purpusii* 'Winter Beauty'
Coronilla valentina subsp. *glauca*	*Mahonia* x *media* cultivars
Crataegus monogyna 'Biflora'	*Narcissus cantabricus*
Cyclamen coum	*Narcissus papyraceus*
Daphne bholua cultivars	*Prunus* x *subhirtella*
Erica carnea cultivar	*Sarcococca confusa*
Erica x *darleyensis* cultivars	*Viburnum* x *bodnantense* cultivars
Fatsia japonica	*Viburnum farreri*
Galanthus spp.	*Viburnum farreri* 'Candidissimum'
Garrya elliptica 'James Roof'	*Viburnum tinus*

JANUARY		
Abeliophyllum distichum	*Cyclamen coum*	*Iris reticulata*
Acacia dealbata	*Daphne bholua* cultivars	*Iris unguicularis*
Azara microphylla	*Daphne laureola* subsp. *philippi*	*Jasinum nudiflorum*
Camellia japonica cultivars	*Daphne orora* 'Aureomarginata'	*Lonicera fragrantissima*
Camellia sasanqua cultivars	*Drimys winteri*	*Mahonia* x *media* cultivar
Camellia williamsii hybrids	*Eranthis hyemalis*	*Narcissus cantabricus*
Chaenomeles japonica cultivars	*Erica carnea* cultivars	*Narcissus papyraceus*
Chaenomeles speciosa cultivars	*Erica* x *darleyensis* cultivars	*Parrotia persica*
Chaenomeles x *superba* cultivars	*Galanthus* spp.	*Prunus* x *subhirtella*
Chimonanthus praecox	*Garrya elliptica* 'James Roof'	*Rhododendron mucronulatum*
Clematis cirrhosa	Grasses	*Ribes laurifolium*
Clematis napaulensis	*Hamamelis* spp. and cultivars	*Sarcococca confusa*
Clematis urophylla 'Winter Beauty'	*Helleborus argutifolius*	*Sarcococca hookeriana* var. *digyna*
Coronilla valentina subsp. *glauca*	*Helleborus niger*	*Viburnum* x *bodnantense* cultivar
Corylus avellana 'Contorta'	*Helleborus* 'Walberton's Rosemary'	*Viburnum farreri*
Crataegus monogyna 'Biflora'	*Helleborus* x *hybridus*	*Viburnum farreri* 'Candidissimum'
Crocus tommasinianus	*Hydrangea febrifuga*	*Viburnum tinus*

FEBRUARY	
Abeliophyllum distichum	*Helleborus argutifolius*
Acacia dealbata	*Helleborus niger*
Azara microphylla	*Helleborus* 'Walberton's Rosemary'
Camellia japonica cultivars	*Helleborus* x *hybridus*
Camellia williamsii hybrids	*Iris reticulata*
Celtis australis	*Iris unguicularis*
Chaenomeles japonica cultivars	*Jasminum nudiflorum*
Chaenomeles speciosa cultivars	*Leucojum vernum*
Chaenomeles x *superba* cultivars	*Lonicera fragrantissima*
Chrysosplenium macrophyllum	*Lonicera standishii*
Clematis armandii	*Mahonia* x *media* cultivars
Clematis cirrhosa	*Narcissus cantabricus*
Clematis napaulensis	*Narcissus cyclemineus*
Clematis urophylla 'Winter Beauty'	*Narcissus* 'February Gold'
Cornus mas	*Narcissus papyraceus*
Coronilla valentina subsp. *glauca*	*Narcissus psuedonarcissus*
Corylus avellana 'Contorta'	*Primula vulgaris*
Crataegus monogyna 'Biflora'	*Prunus incisa* 'Praecox'
Crocus chrysanthus	*Prunus mume* 'Beni-chidori'
Crocus tommasinianus	*Prunus* x *subhirtella*
Cyclamen coum	*Pulmonaria rubra*
Daphne bholua cultivars	*Puschkinia scilloides* var. *libanotica*
Daphne laureola	*Rhododendron mucronulatum*
Daphne laureola subsp. *philippi*	*Ribes laurifolium*
Daphne orora 'Aureomarginata'	*Sarcococca confusa*
Distylium racemosum	*Sarcococca hookeriana* var. *digyna*
Edgeworthia chrysantha	*Skimmia* x *confusa* 'Kew Green'
Eranthis hyemalis	*Sophora micropyhlla* 'Sun King'
Erica carnea cultivars	*Stachyurus preacox*
Erica x *darleyensis* cultivars	*Tropaeolum tricolor*
Euphorbia characias subsp. *wulfenii*	*Viburnum* x *bodantense* cultivars
Galanthus spp.	*Viburnum farreri*
Garrya elliptica 'James Roof'	*Viburnum farreri* 'Candidissimum'
Grasses	*Viburnum tinus*
Hamamelis spp. and cultivars	*Vinca minor* 'Alba variegata'

***Narcissus* 'February Gold'**

Helleborus* x *hybridus

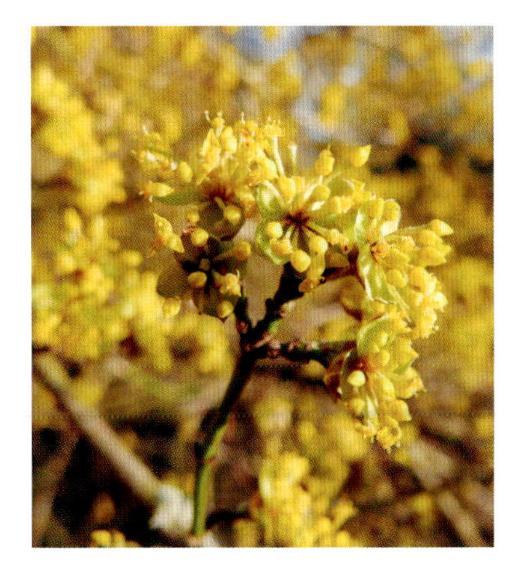

Cornus mas

Flower colours

RED	
Camellia sasanqua 'Red'	*Hamamelis vernalis* 'Red Imp'
Chaenomeles japonica	*Hamamelis* x *intermedia* 'Diane'
Chaenomeles japonica 'Sargentii'	*Helleborus* x *hybridus*
Chaenomeles x *superba* 'Crimson and Gold'	*Hamamelis* x *intermedia* 'Jelena'
Corylus avellana 'Red Majestic'	*Pulmonaria rubra*
Distylium racemosum	*Skimmia japonica* 'Rubella'
Edgeworthia chrysantha	*Tropaeolum tricolor*
Erica carnea 'Winter Rubin'	

GREEN
Chaenomeles speciosa 'Kinshiden'
Chaenomeles x *superba* 'Lemon and Lime'
Daphne laureola subsp. *philippi*
Euphorbia characias subsp. *wulfenii*
Helleborus argutifolius
Helleborus x *hybridus*
Snowdrop cultivars

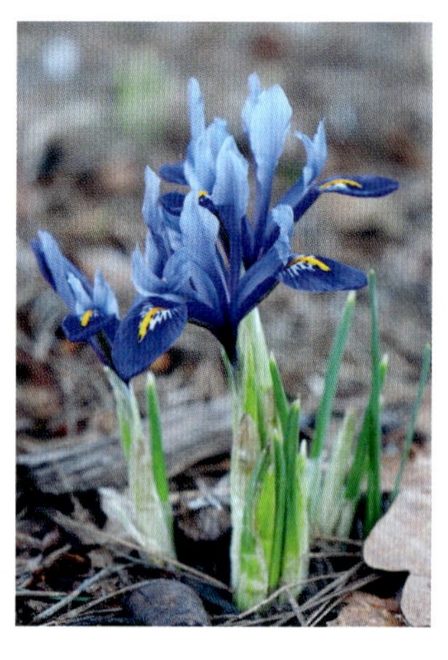

***Iris reticulata* 'Harmony'**

BLUE
Iris reticulata
Iris reticulata 'Fabiola'
Iris reticulata 'Harmony'
Iris unguicularis
Puschkinia scilloides var. *libontica*

ORANGE
Chaenomeles japonica 'Cido'
Chaenomeles japonica 'Orange Beauty'
Hamamelis x *intermedia* 'Orange Peel'

PURPLE

Crocus tommasinianus

Cyclamen coum

Erica carnea 'December Red'

Helleborus x *hybridus*

Iris reticulata 'Pauline'

Rhododendron mucronulatum

***Erica carnea* 'Winter Rubin'**

Leucojum vernum

PINK	
Camellia 'Bow Bells'	*Erica* x *darleyensis* 'Furzey'
Camellia 'St Ewe'	*Erica* x *darleyensis* 'Phoebe'
Camellia japonica 'Devonia'	*Erica* x *darleyensis* 'Tweety'
Camellia japonica 'Lady Vansittart'	*Helleborus* 'Walberton's Rosemary'
Camellia sasanqua 'Peach Blossom'	*Helleborus* x *hybridus*
Chaenomeles speciosa 'Madame Butterfly'	*Prunus cerasifera* 'Pissardii'
Chaenomeles x *superba* 'Cameo'	*Prunus mume* 'Beni-chidori'
Cyclamen coum	*Prunus* x *subhirtella*
Daphne bholua 'Darjeeling'	*Viburnum farreri*
Daphne bholua 'Jacqueline Postill'	*Viburnum tinus* 'Eve Price'
Daphne odora 'Aureomarginata'	*Viburnum* x *bodnantense* 'Charles Lamont'

WHITE	
Abeliophyllum distichum	*Helleborus* x *hybridus*
Camellia sasanqua 'Mine-no-yuki'	*Iris reticulata* 'Katharine Hodgkin'
Chaenomeles speciosa 'Nivalis'	*Leucojum vernum*
Chimonanthus praecox	*Lonicera fragrantissima*
Chrysosplenium macrophyllum	*Lonicera standishii*
Clematis armandii	*Narcissus cantabricus*
Clematis cirrhosa	*Narcissus papyraceus*
Clematis napaulensis	*Prunus incisa* 'Praecox'
Clematis urophylla 'Winter Beauty'	*Puschkinia scilloides* var. *libanotica* 'Alba'
Crataegus monogyna 'Biflora'	*Sarcococca confusa*
Daphne bholua 'Alba'	*Sarcococca hookeriana* var. *digyna*
Daphne bholua 'Gurkha'	*Skimmia* x *confusa* 'Kew Green'
Erica carnea 'Snowbelle'	*Viburnum farreri* 'Candidissimum'
Erica x *darleyensis* 'White Perfection'	*Viburnum tinus* 'Lisarose'
Galanthus spp.	*Viburnum* x *bodnantense* 'Dawn'
Helleborus niger	*Vinca minor* 'Alba Aureovariegata'

YELLOW	
Acacia dealbata	*Jasminum nudiflorum*
Azara microphylla	*Mahonia* x *media* 'Charity'
Camellia 'Jury's Yellow'	*Mahonia* x *media* 'Winter Sun'
Chimonanthus praecox 'Luteus'	*Narcissus cylemineus*
Cornus mas	*Narcissus* 'February Gold'
Crocus chrysanthus	*Narcissus* 'February Silver'
Edgeworthia chrysantha	*Narcissus psuedonarcissus*
Eranthis hyemalis	*Narcissus* 'Rijnveld's Early Sensation'
Hamamelis mollis 'Brevipetala'	*Primula vulgaris*
Hamamelis vernalis 'Squib'	*Ribes laurifolium*
Hamamelis x *intermedia* 'Angelly'	*Sophora microphylla* 'Sun King'
Helleborus x *hybridus*	*Stachyurus praecox*

Fragrant plants

FLOWERS	
Bulbs	**Shrubs**
Crocus chrysanthus	*Abeliophyllum distichum*
Galanthus nivalis	*Azara microphylla*
Iris reticulata	*Chimonanthus praecox*
Iris unguicularis	*Coronilla valentina* subsp. *glauca*
Narcissus papyraceus	*Daphne bholua*
	Daphne laureola subsp. *philippi*
Climbers	*Daphne odora*
Clematis armandii	*Edgeworthia chrysantha*
Clematis cirrhosa	*Hamamelis* spp. and cultivars
	Lonicera fragrantissima
Trees	*Lonicera standishii*
Acacia dealbata	*Mahonia* x *media*
Azara microphylla	*Sarcococca confusa*
Prunus mume 'Beni-chidori'	*Sarcococca hookeriana* var. *digyna*
	Skimmia x *confusa* 'Kew Green'
Perennials	*Viburnum farreri*
Primula vulgaris	*Viburnum farreri* 'Candidissimum'
	Viburnum x *bodnantense* and cultivars

Crocus chrysanthus

Sarcococca confusa

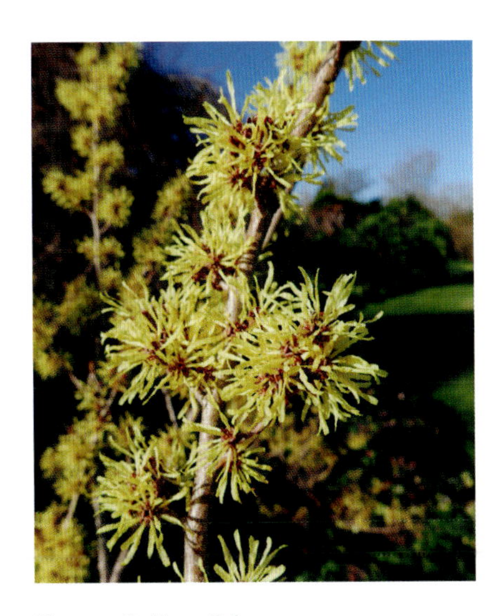

Hamamelis 'Angelly'

OPPOSITE
Clematis cirrhosa

Daphne bholua

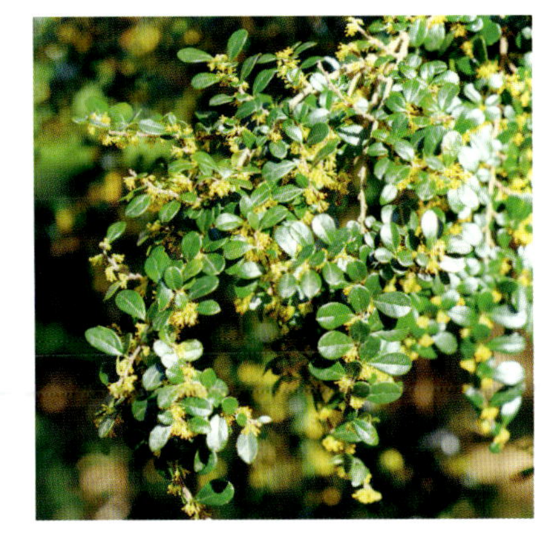

Azara microphylla

Plants for wildlife

Bulbs	**Shrubs**		**Trees**
Crocus chrysanthus – Bees	*Callicarpa giraldii* – Birds	*Hamamelis* spp. and cultivars – Bees and other pollinators	*Cornus mas* – Bees
Crocus tommasinianus – Bees	*Chaenomeles japonica* – Bees	*Lonicera fragrantissima* – Bees	*Crataegus monogyna* 'Biflora' – Bees
Galanthus elwesii – Bees	*Chaenomeles speciosa* – Bees	*Lonicera standishii* – Bees	*Ilex* spp. and cultivars – Bees and birds
Galanthus nivalis – Bees	*Chaenomeles* x *superba* – Bees	*Mahonia* x *media* – Bees	*Prunus incisa* 'Praecox' – Bees
Galanthus plicatus – Bees	*Coronilla valentina* subsp. *glauca* – Bees	*Pyracantha* cultivars – Birds	*Prunus mume* 'Beni-chidori' – Bees
Iris reticulata – Bees	*Cotoneaster lacteus* – Birds	*Sarcococca confusa* – Bees	*Prunus* x *subhirtella* – Bees
Puschkinia scilloides var. *libontica* – Bees	*Daphne bholua* - Bees	*Sarcococca hookeriana* var. *digyna* – Bees	
	Erica carnea – Bees	*Stachyurus praecox* – Bees	**Perennials**
Climbers	*Erica* x *darleyensis* – Bees	*Symphoricarpos albus* – Birds	*Euphorbia characias* subsp. *wulfenii* – Bees
Clematis cirrhosa – Bees	*Fatshedera lizei* – Bees and birds	*Viburnum tinus* – Bees	*Helleborus argutifolius* – Bees
Hedera spp. – Bees, birds, butterflies, moths & others	*Fatsia japonica* – Bees and birds		*Helleborus niger* – Bees
			Helleborus 'Walberton's Rosemary' – Bees
			Helleborus x *hybridus* – Bees
			Primula vulgaris – Bees and other pollinators
			Pulmonaria rubra – Bees

CLOCKWISE FROM TOP LEFT
Redwing, winter visitor feeding on various berries; red admiral, surprise winter visitor; honey bee on *Lonicera*; and hoverfly on *Viburnum* flowers

Foliage and stem colour

Cornus and *Rubus* stems

Fatshedera lizei 'Annemieke'

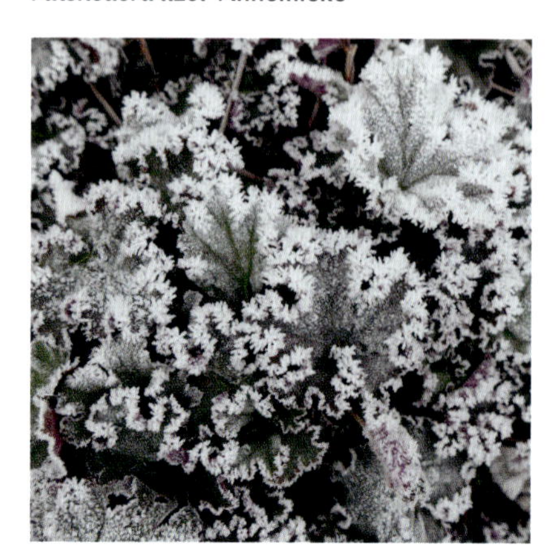

Frosted *Heuchera*

Climbers	**Shrubs**
Hedera spp. and cultivars	*Aucuba japonica* 'Variegata'
Trachelospermum asiaticum 'Summer Sunset'	*Choisya ternata* 'Sundance'
Trachelospermum jasminoides 'Variegatum'	*Cornus alba* 'Siberica'
	Cornus alba 'Kesselringii'
Trees	*Cornus alba* 'Westonbirt'
Acer davidii	*Cornus sanguinea* 'Midwinter Fire'
Acer griseum	*Cornus sanguinea* 'Anny's Winter Orange'
Arbutus x *andrachnoides*	*Cornus sericea* 'Bud's Yellow'
Betula utilis subsp. *albosinensis* 'Fascination'	*Cornus sericea* 'Flaviramea'
Betula utils subsp. *jacquemontii* 'Doorenbos'	*Elaeagnus pungens* 'Maculata'
Eucalyptus pauciflora subsp. *niphophila*	*Euonymus fortunei* 'Emerald Gaiety'
Pinus bungeana	*Euonymus fortunei* 'Emerald 'n' Gold'
Pinus spp.	*Euonymus fortunei* 'Silver Queen'
Prunus serrula	*Fatsia japonica*
Salix alba subsp. *vitellina* 'Britzensis'	*Heptacodium miconioides*
Salix alba subsp. *vitellina* 'Yelverton'	*Nandina domestica*
Tilia cordata 'Winter Orange'	*Pittosporum* spp.
	Rhododendron 'President Roosevelt'
Perennial	*Ribes biflorus*
Ajuga reptans	*Rubus cockburnianus*
Bergenia purpurascens	*Rubus cockburnianus* 'Goldenvale'
Heuchera	*Rubus thibetanus*
Ophiopogon planiscapus 'Black Beard'	*Salix alba* subsp. *vitellina* 'Yelverton'
Pachysandra terminalis 'Variegata'	x *Fatshedera lizei*

OPPOSITE
Ivy 'Gold Heart'

Index

Entries with a photograph are indicated by a **bold** page number.

OPPOSITE
A carpet of frost covered black leaves beneath white-stemmed birches

OPPOSITE
Close-up of the pretty, fragrant flowers of wintersweet (*Chimonanthus*)

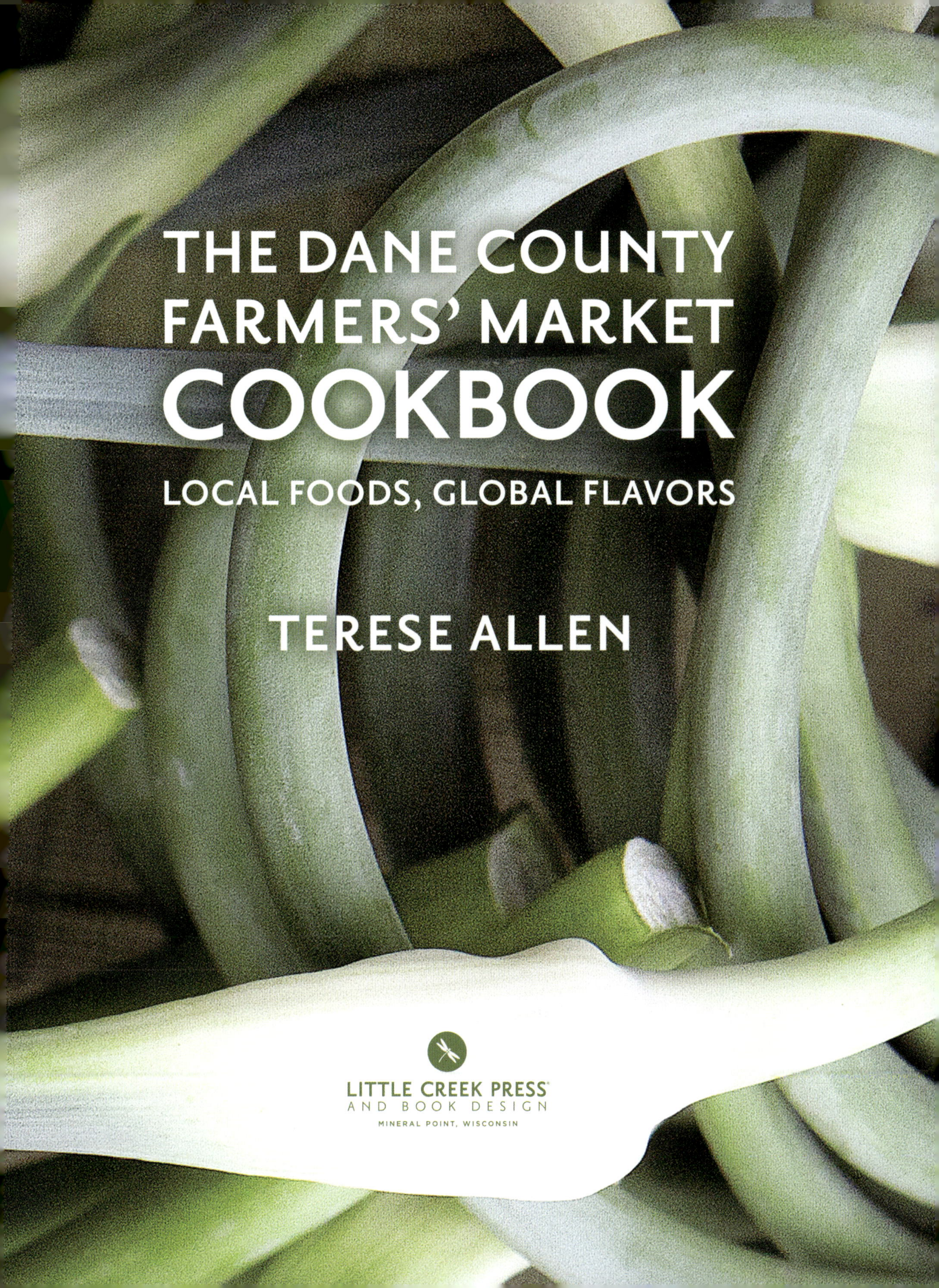
THE DANE COUNTY
FARMERS' MARKET
COOKBOOK
LOCAL FOODS, GLOBAL FLAVORS
TERESE ALLEN
LITTLE CREEK PRESS
AND BOOK DESIGN
MINERAL POINT, WISCONSIN

PRAISE FOR *THE DANE COUNTY FARMERS' MARKET COOKBOOK*

"The Midwest is lucky to claim Terese Allen as keeper of our culinary heritage and advocate for locally grown, seasonal cooking and eating. Now, in this eagerly awaited DCFM cookbook, Terese takes us on a foodie's tour of that beloved market as only she can, warmly welcoming us in to the circle of growers, makers, chefs and home cooks who for 50 years have been transforming how we cook, eat and think about food. In sharing their stories and favorite recipes, she enriches our appreciation of our region's diversity and expands our definition of 'community.' In the intro, Terese refers to DCFM as 'a unique merger of taste and place'—a perfect description of this remarkable book!"

—Kate Thompson, Director, Wisconsin Historical Society Press

"The DCFM made a brilliant choice asking Terese Allen to create this community cookbook. The history of the market, the people, the region and the food are richly detailed."

—Catherine Lambrecht, Greater Midwest Foodways Alliance

"The only thing I like more than running into Terese Allen at the fruit stand near dawn is diving into her capacious, diverse exploration of the beauties of the Dane County Farmers' Market. She captures the ever-changing array of growing and cooking traditions with the joy and care befitting this beloved gathering spot for cooks, eaters, farmers. Now how am I supposed to do anything but plan my next meal?"

—Michelle Wildgen, author of *Wine People, Bread and Butter,* and *You're Not You*

"Terese Allen has shared the joy and power of the DCFM, a gathering space that cultivates love for our food systems to grow beyond the Capitol Square ... and makes me even prouder to be from and to represent Wisconsin."

—Francesca Hong, Wisconsin State Representative; chef and co-owner of Morris Ramen

"Anyone who has visited the DCFM or eaten food sourced from it will tell you: This market is our community pride and joy, and a fount of health and economic vitality. Over decades entire platoons of home cooks and eaters have joined farmers and chefs on the front lines of this market to wage a happy concord of wellness, a path that wends from soil to plants to humans to communities to planet. Terese Allen, herself a proud culinary daughter of this market, translates the local-global vitality of its many diverse vendors and provender into understandable recipes, each a microcosm of the market's creativity, resourcefulness and deliciousness. It's all here: a 50-year perspective, engaging stories and useful kitchen tips."

—Odessa Piper, James Beard Award winner and founder of L'Etoile restaurant

"Talk about a masterpiece on 'soul food'! That's what this book is, covering the very soul of one of the most revered farmers markets in the nation. With its luscious buffet of recipes, photos and stories, it's the next best thing to being there."

—Scott Warner, president, Culinary Historians of Chicago

"The Dane County Farmers' Market shines as a beacon for those wishing to experience Wisconsin's nuanced agricultural heritage, our culinary prowess and our diversity. Over time the DCFM has impelled countless cooks and food enthusiasts like Terese Allen to develop the network between themselves and their food that creates COMMUNITY. In this book, Terese introduces the ingredients, recipes and stories that invite us to fall in love with the market and the food of the Upper Midwest all over again, and to truly live the connection between land, people and table."

—Luke Zahm, host of PBS's "Wisconsin Foodie" and chef-owner of Driftless Café

"This engaging cookbook is like the best trips to the farmers market—full of color, variety and lots of friendly faces. The recipes are surprising and accessible, sourced from vendors, chefs and market-goers alike. I was drawn to the kimchi fried rice and fried green tomatoes, German kohlrabi, Hmong pork, Nepalese curry and Wisconsin cheese soup, plus so much more. For anyone who wants to cook more seasonally, this book is a treasure—and a vivid love letter to Madison, too."

—Lindsay Christians, author of *Madison Chefs: Stories of Food, Farms and People* and *The Osteria Papavero Cookbook*

"I vividly recall that first Dane County Farmers' Market on a chilly fall morning in 1972. I worried there might be a poor start, but the notion of bringing city and country folk together with fresh produce on the Square was embraced by all. One older farmer told me, "Don't worry, if the fruit is ripe you don't have to shake the tree so hard." He was right. And now, more than 50 years later, we have this beautiful, excellent book to celebrate the market."

—Jonathan Barry, DCFM's founding manager and former Dane County Executive

"Ace food historian-activist-chef Terese Allen introduces Madison's beloved farmers market for what it is: an extraordinary array of seasonal and sometimes-unusual ingredients that need not baffle the average home cook. Expect a melting pot of international recipes from cooks as local as the market vendors. In the mix: culinary challenges for when we're motivated and easy-make recipes for when we're not."

—Mary Bergin, author of *Wisconsin Supper Club Cookbook* and *Small-Town Wisconsin*

"The Dane County Farmers' Market Cookbook is truly a gift to lovers and supporters of farmers markets, and for anyone who appreciates and knows the incredibly hard work that goes into growing and producing fresh local food. Sit back, relax and take a trip through the farmers market with an enjoyable read—what a treat for food lovers everywhere!"

—Douglas Merriam, author-publisher of *Farm Fresh Journey: The Santa Fe Farmers Market Cookbook*

"The Dane County Farmers' Market Cookbook doesn't just reflect the market's tangible vibe; it mirrors the interests and intentionality of the community that flocks there to shop. The recipes here are both of the moment and timeless, both specific and universal, and they come from the kitchen of every-cook, who's just trying to make a delicious, healthy and yet stress-free meal."

—Linda Falkenstein, Associate Editor, Isthmus newspaper

"The DCFM cookbook is not just a cookbook, but brings to life the history and importance of the market—you can't help feeling proud and a bit nostalgic. I love the international cuisine emphasis to introduce new foods to consumers."

—Philip Kauth, Exec. Director, REAP Food Group

Little Creek Press
5341 Sunny Ridge Road
Mineral Point, WI 53565

TO ORDER BOOKS: www.littlecreekpress.com or email: info@littlecreekpress.com

Printed in Wisconsin, United States of America

Cataloging-in-Publication Data
Names: Allen, Terese, author
Title: The Dane County Farmers' Market Cookbook: Local Foods, Global Flavors
Description: Mineral Point, WI. Little Creek Press, 2023
Identifiers: LCCN: 2023909160 | ISBN: 978-1-955656-51-1
Subjects:
COOKING / Farm to Table
COOKING / Regional & Ethnic-General
COOKING / Seasonal

Book design by Little Creek Press with help from Lily Stern

All photographs are by William C. Lubing unless otherwise indicated

Front and back cover photo credits:
Front cover photo: Lucinda Ranney
Back cover photos: farmstand, Lucinda Ranney; mushroom grower, Lucinda Ranney; child, Lois Bergerson; cheese, William C. Lubing; stew, Zainab Hassen; peppers, William C. Lubing

All proceeds benefit the Dane County Farmers' Market.

In gratitude to growers and cooks

And in joyful memory of market lover
Kathy Trudell

Recipe on page 110
Meri Tunison

TABLE OF CONTENTS

FOREWORD

BY TORY MILLER, JAMES BEARD AWARD WINNER AND CO-OWNER OF L'ETOILE AND GRAZE RESTAURANTS

Ruthie Hauge (The Cap Times)

I arrived in Madison in 2003, after stints as a culinary student, a restaurant cook and eventually a sous chef in New York City. It was a time before the internet, when the world felt a little smaller than it does now. I didn't know much about Wisconsin's capital city, but even in Manhattan the Dane County Farmers' Market (DCFM) was a legend. When I was considering a move to Madison, several chefs and colleagues I had worked with advised me to look up "this chef named Odessa" and told me there was a good farmers market there. This sounded exciting to me at the time, but as I think back on it now, it's astounding that line cooks in the Big Apple would be talking about a farmers market in Wisconsin. They didn't even know where Wisconsin is. I actually would have to explain to them: "You know where Chicago is? Well, go straight up from there and you'll be in Wisconsin."

My first stop upon arrival in Madison was the kitchen at L'Etoile, which at the time was one of the very few restaurants sourcing the majority of their ingredients directly from farmers and the farmers market. These days, however, I have to get up earlier and earlier to race other chefs around the DCFM at the Capitol Square, a "chore" I am happy to do, as I am so glad about how much the grower-chef connection has grown since back then.

My relationship with the DCFM began on the first Saturday of October in 2003, when I pulled a little red wagon out of the restaurant to shop with Chef Odessa Piper. It started out as you might think: A chef wants to find the best ingredients. The best ingredients are grown by local farmers. The best place to find farmers is a market. And at L'Etoile, we happened to have the largest producers-only market in the country right outside our door.

Suffice it to say I was hooked. So for the last twenty years I've engineered my schedule around weekly trips to the Square on Saturday mornings. And now, going out to market is perhaps both the most important part of my job and, arguably, of the life that I have built in Madison.

The DCFM changed the game for me. Yes, it's because of the huge variety of incredible produce available at the market throughout the Wisconsin seasons. Yes, it's a playground for a chef out there—table after table full of the freshest, most colorful, most delicious vegetables, fruits, herbs, cheeses and meats. And I wanted all of it. But what struck me the most—and what led me to fall in love with Madison and build my home there—were the people behind those tables.

The relationship between chefs and farmers goes beyond the foodstuffs that growers provide and chefs utilize. It is a working relationship, so obviously there is money involved, and at the end of the day, we all have to make it. But the market makes the relationship something more. Every week there is a new story, a new struggle, a new victory, and often a new product to try. We share our lives with one another, week after week. I get a genuine sense of caring from the farmers, not just about the food they are selling, but who they are selling it to, and what will be created with it. Inspiration abounds in those four sides of the Capitol Square. For a chef, rushing back to the kitchen with a market haul and cooking with it is addicting. Still, I often find myself inspired as much by the people growing our food as the ingredients themselves.

I have watched as the market continues to evolve. There's a new generation of producers who have joined old-school growers, and a large population of curious cooks and accomplished chefs who head out every week to see what the market has in store for us. One of the people who is always at the market (often before me!) is Terese Allen. A longtime chronicler of Wisconsin's culinary history and culture, Terese has also been an influential supporter of local foods since before I was anywhere near this incredible food scene. In this cookbook, she captures the vivid character of the DCFM and shows us how its ever-growing array of local ingredients can be the basis for globally inspired meals.

She also shines a light on the amazing people who make up the market. As you read their stories and explore their recipes, you'll really get to know the community that has grown to define the market itself. We are farmers, producers, chefs and home cooks, all of us sharing our love for the best ingredients—and something delicious to make out of them.

So grab this book, get to the market, and we'll see you there.

ONE WAY
BRUNKOW CHEESE

INTRODUCTION

Back in the early 1970s, when a handful of forward thinkers were looking to start a farmers market in Madison, the obstacles must have seemed daunting. It had been more than a half-century since a market had operated in the city. The organizers needed to find a location. They had to round up vendors. There were permits, health codes and insurance to deal with, people in power to convince. And who knew if any customers would even show up?

But if those early boosters were worried, they needn't have been. They launched the Dane County Farmers' Market on September 30, 1972. It grew rapidly into one of the largest and most acclaimed farmers markets in the nation—the market that revitalized downtown Madison, helped area agriculture flourish and changed the character of food and cuisine in the region.

Just as cheese, beer and brats come to mind when you hear "Wisconsin," so does the mention of Madison conjure the Dane County Farmers' Market. The city's most celebrated event takes place on Saturday mornings from April to November beneath the majestic, tree-framed State Capitol dome. Shoppers stroll counterclockwise around the Capitol Square, perusing stands, tasting samples and filling sacks and wagons with seasonal bounty. Vendors—up to 150 of them at a time filling available spots in high season—tout a staggering array of Wisconsin-grown produce, meats, cheeses, bakery goods, beverages and specialty products.

On Wednesdays a second market operates on Martin Luther King Jr. Boulevard between the stately government buildings located down the street from the Capitol. During the cold months of the year, the DCFM moves indoors into two of the city's architectural gems: the lake-hugging, Frank Lloyd Wright-designed Monona Terrace Community and Convention Center (for the Holiday Markets), and Garver Feed Mill, a handsomely renovated turn-of-the-century event complex (for the Winter Markets).

Indoors or outdoors, the DCFM is a unique merger of "taste" and place, a culinary and cultural draw for shoppers, home cooks, chefs, tourists, families, breakfast lovers, photographers, foodies, University of Wisconsin students—anyone who eats, really. You don't need to be a locavore to enjoy or benefit from this market, but if you are one, the DCFM will pretty much blow your mind. I've been attending it for more than 40 years myself, and it still blows mine.

The evolution of the Dane County Farmers' Market can be traced in the crops and products that have been sold there. (For a history of the market, see "The First Fifty Years," pages 234–243.) During its early days the offerings were worker-bee vegetables like potatoes, carrots, beans, corn and tomatoes, foodstuffs that drew crowds because they were, well, fresh—that is to say, so much better than most of their counterparts at the grocery store. It didn't take long, though, for more unusual ingredients to hit the scene.

Heirloom tomatoes, goat cheese, ready-to-eat pesto—these and many more items made a splash, and then, in turn, became regular purchases on Saturday mornings. Chefs lent cachet to specialties like fava beans and Jerusalem artichokes by putting them on their menus (and sometimes causing a stampede on available supplies). Hmong growers introduced vibrant new tastes to the line-up—lemongrass, squash blossoms, Thai eggplants. Young farmers brought in microgreens and cippolini onions; oldsters taught us about Concord grapes, elderflower berries and hickory nuts. These days, the number of items available at the market is nearly uncountable. (You'll see an informal attempt at this, however, on pages 245–247.)

The city's most celebrated event takes place on Saturday mornings from April to November beneath the majestic, tree-framed State Capitol dome.

Clearly, it's not just about potatoes and carrots anymore. There are so many choices that shoppers no longer need to look elsewhere as much to supply their pantries or complete their menus. The market boasts all the basics and an enormity of extras. Fresh ginger, dried curry leaves, bitter melon. Gluten-free bread, pear cider, black garlic. Focaccia, sweet potatoes, epazote. Harissa paste, guinea hens, duck eggs.

This book celebrates the growth, progress and influence of the DCFM—how it started with a handful of farmers selling staples and expanded over a half-century to hundreds of vendors, tens of thousands of weekly customers, and an ever-growing variety of crops, world-class cheeses, pastured meats and specialty products.

During her tenure as manager of the DCFM (2016–2021), Sarah Elliott often talked about the market as "a place to do your weekly grocery shopping." In an interview for this book, she told me, "We always tried to remind people that you can find the 'whole plate' at market—

fruits and veggies, meat, cheese, breads and pastries, and also maple syrup, honey and so much more."

This cookbook is about what to do with all those goodies. Drawing from a world of local ingredients, its recipes are inspired by cuisines from around the globe. Come join us—the members of the market community—in a salute to the Dane County Farmers' Market.

Chef Tory Miller, in the Foreword to this book, tells us that the DCFM changed the game for him. It did for me, too ... about twenty years earlier, that is. This was during the 1980s, a decadent decade—a time when oversized plates held undersized portions of expensive, fussed-over food, when culinary gadgets took over our countertops, and cooks labored to master the art of epicurean cuisine.

Me, I was a budding chef at the time. A few years out of cooking school, I ran the kitchen of one of the Ovens of Brittany restaurants, known then in the Madison region for such elaborate presentations as spinach mushroom gateau and chocolate Queen of Sheba torte. I looked up to people who spent their discretionary income on multi-course meals and pricey wines, who subscribed to trendy cooking magazines and knew their latte from their lassi.

Funny thing, though. While I could down raw oysters with the best of them and prided myself on my crepe-flipping technique, the food that made me the happiest wasn't costly or gourmet. What really blissed me out was going to the Capitol Square on Saturday mornings and then cooking with the colorful, in-season ingredients I found there.

I'd get to the market early, before the restaurant opened, to see what called out to me: curious softball-size muskmelons and voluptuous blueberries, farm-raised rainbow trout and bags of fragrant herbs. Back in the kitchen, the purchases became brunch and dinner specials: mini-melon halves filled with juicy berries and honey-sweetened yogurt; trout stuffed with organic leeks and carrots; fettuccine with fresh herbs and cream. These simple dishes were tasty and well-received, and I found more satisfaction in cooking them than any of the meals that required imported ingredients or tricked-out sleights of hand.

Although I didn't know it at the time, I was joining a food revolution. I was drawn first to its deliciousness—and, I'll admit, to that happy look people gave me when they tasted local food at its peak-season best. But I also quickly succumbed to other pleasures and benefits of farm-to-table fare: how it is health-giving, how it sustains the land, supports small farmers, improves regional economies, builds human understanding and connection, and weaves urban and rural communities together.

Once I began attending the DCFM regularly, there was no going back. My cooking transformed, friendships grew, and the market simply became a part of me. It became—and remains—touchstone and muse, a weekly ritual and a source of joy. There are innumerable others who have similar feelings. In the sidebar on this and the following page, and throughout this book, you'll hear from recipe contributors and other DCFM regulars about why they go to market, what it means to them, too.

What's In It for You

This collection took root one morning early in 2022. I had stopped at Ted and Joan Ballweg's Savory Accents stand at Garver Feed Mill, the DCFM's indoor Winter Market venue. I knew

THE MARKET SPEAKS FOR ITSELF

"The DCFM is such a wonderful place to explore and be inspired by the agricultural diversity of Wisconsin. I have many fond memories over the years of finding things there that helped grow my appetite for gardening and cooking as well as my appreciation for the skilled farmers and food makers of this region. Every visit opens up a whole world of new possibilities!"

—Emily Rose Haga

"Agriculture is inherently a risky business. Small family farmers ... must meet the rising demand for more food of higher quality while combating weather volatility, equipment and land cost investment, product marketing struggles, labor shortages and ever-changing customer tastes and expectations. DCFM has been a valuable tool for us to sell our product and connect directly with our customers and fellow growers. My hope is DCFM will always keep their focus and support for the small local farmer, helping us to discover and expand our farming options."

—Julie Sutter, Sutter's Ridge Farm (excerpted from the Dane County Farmers' Market Newsletter, October 5, 2022)

"I like that my meals are made of the produce that is in season, which brings natural variation to the dishes I make. It is always a nice challenge to think about what dishes could work with what I could get at the market. I also love that I am supporting local businesses [by] buying at the market. Most importantly, the fresh vegetables from the market actually do taste like vegetables and not just like water, as [with] the vegetables from the store."

—Natascha Merten

something was up because of the unusually bright smiles that the two longtime pepper vendors were shooting my way, looks that said, "This isn't going to be just our usual Saturday chat." Pulling me over, they told me the market was taking proposals for projects to celebrate its 50th anniversary year. They asked, had I ever thought about doing a Dane County Farmers' Market cookbook?

I had, as a matter of fact. That was back in the early 1990s, during the development of my second cookbook, which started out with a focus on the DCFM and then morphed into an exploration of farmers markets around the state. I became an all-out farmers market devotee with that book, someone who plans pretty much everything around them—my day, my week, vacations, life. Still, Madison's market was my first love—the weekly event that had shaped my

"For many makers and farmers, the work during the week is solitary. On a market Saturday, while exhausting, it's also very rewarding to interact with thousands of people who show up to support their local food economy. Having the chance to interact with customers, get feedback, share stories, answer questions and meet their families makes our work all the more rewarding."

—Kirk Smock, Origin Breads (excerpted from Dane County Farmers' Market Newsletter, May 11, 2022)

"[I am an] enthusiastic and grateful customer of the DCFM [and] the smaller markets that complement the Capitol one. It's a wonderful resource for anyone who prefers local, seasonal food and enjoys meeting and getting to know the local farmers who care for the soil and provide such a variety of life-sustaining foodstuffs."

—Sabine Gross

"I like to go to look for fun cheeses, short-season vegetables like sorrel, peppers, heirloom apples ... and surprises. There is always a surprise."

—Linda Falkenstein

"Hands down, it is the finest farmers market I have ever visited, anywhere, ever."

—James Wehn

culinary outlook—and as is the case with all first loves, I had an imprinted fidelity to it. When the Ballwegs suggested we form a committee and produce a cookbook together, my reaction was a slam-dunk "Yes."

It didn't take long for us to choose a theme. Ted's idea to feature market ingredients in multicultural dishes lit us up, for it dovetailed perfectly with the anniversary's focus on the DCFM's growth and progress over its lifespan. What better way to showcase the huge range of regionally grown foodstuffs available at the market than to feature them in recipes from around the world? What better way to celebrate the diversity and interdependence of the market community itself?

All the recipes in this book came from members of that community—vendors, shoppers, cooks, chefs, volunteers and staff. When asked for submissions that highlight both local ingredients and international flavors, they responded with more than twice what we had room for, leaving me with the sometimes daunting, always intriguing task of curating recipes for the book. We wanted to underscore the diversity of goods at the market and to represent as many areas of the world as possible. No book would have enough room for them all, of course, but I hope you'll agree that this one proves how doable and appetizing it can be to "shop local and cook global."

The recipes spring from many sources, with the most heartfelt ones coming from family and community, and from ethnic, regional and national traditions. Vendor Phil Yang's eyes lit up when I asked him about his mother's mustard greens with boiling pork, a favorite of his that epitomizes the simplicity, healthfulness and savor of Hmong cooking. Shopper Natascha Merten shared a nostalgic, passed-down-through-the-generations German kohlrabi in *mehlschwitze* (white sauce) served with potatoes and sausages. Oneida Nation member Dan Cornelius honors his ancestors with a contemporary salad that features indigenous wild rice, shell beans and hominy.

Many recipes came with stories. Efrat Livny wrote about growing up and learning to cook in Israel, and how making *shakshuka*—eggs poached in a thick tomato and pepper sauce—still connects her to the bustling markets, street foods and restaurants of her homeland's region. Irina Punguil Bravo described *zeama*, a Moldovan chicken-vegetable soup flavored with fresh dill and lovage, fortified with thin noodles and topped with sour cream and spicy red peppers. Irina told us, "It's a nostalgic taste of summers spent at my grandma's in the village, and my mom's cure for our colds. Once I moved to study in Romania, it was my food request every time I came back to Moldova on my breaks. Now, it is the recipe I make for my daughter regularly." In her home country, zeama is traditionally eaten after wedding celebrations—as a hangover remedy.

Favorite ingredients, too, moved contributors to send in their creations. Chef Odessa Piper, founder of Madison's seminal L'Etoile restaurant, has a passion for minty-sweet anise hyssop that finds her stirring it into buttery dough for Scottish shortbread and scattering it over a Greek-like feta-watermelon salad, among other applications. Jesse Brookstein enjoys the fermented meat stick called landjaeger so much that he and his mother perfected a recipe for Louisiana-style jambalaya that features it, just for this book. (Indeed, Jesse loves landjaeger so much that he also wrote an entire book about it.)

Market members are understandably proud of their wares, and many vendor-contributors showed them off in outstanding dishes such as Cliff and Cari Gonyer's unctuous carnitas (shredded braised pork), Jackie Gennett's garlic scape feta pesto, and Sue Gronholz's Scandinavian-style chilled berry soup.

Lucinda Ranney

All the recipes in this book came from members of that community—vendors, shoppers, cooks, chefs, volunteers and staff.

Home cooks offered the recipes that they turn to when the market overflows at harvest time, like Kristin Korevec's *curtido*, a tart, chile-spiked cabbage slaw from El Salvador. Adventurers took inspiration from their travels, too. Madisonian Mary Karau, who has visited Italy more than fifty times, came up with a winner she calls "Wisconsin Sweet Corn Spaghetti" that salutes both her home state and her favorite foreign nation. Cookbooks, cooking classes, restaurants, magazines, the web and more were additional stimuli for contributors.

While a few of the dishes herein are originals, most were inspired by or adapted from one or more of the sources above. The dishes range in complexity from quick-and-easy to average, plus two that are downright projects. The shortest one was a brainstorm for the grill relayed to me one warm fall morning by Tom Murphy of Murphy Farms: "Foil. Veggies. Small meatballs. Cheese curds." That's it. When I asked Tom if he used any butter or oil with the combo, he raised a brow, eyeballed me and called out, emphatically, "BUTTER." "What about herbs?" I added. "Could be!" was the comeback. That's all I could get out of him about the dish, but I immediately wanted to go home and try it.

The two most elaborate recipes came from professionals and are the kinds of challenges that chefs and passionate home cooks love to tackle. Restaurateur Gil Altschul's spunky cocktail, named "This Is a Spicy Drink" and inspired by a chile variety he discovered at the market, starts by chopping up several lemon drop chiles and steeping them in vodka for a week. After that, he juices cucumbers and limes, makes a fruity gum syrup so laborious that even

Gil advises readers to buy it instead, mixes precise amounts of the aforementioned with Peruvian brandy and curaçao, and finally pours the concoction over ice. You can order it at his bar, Gib's, in Madison, or you can follow Gil's instructions and tuck a culinary achievement under your belt.

Grant Johnston's yeast-raised bread is a three-page recipe sure to turn the heads of devoted bakers and lovers of home-baked bread. (In other words, everyone.) Grant is a retired craft beer brewer who knows fermentation like bees know honey. His creation features corn cut fresh off the cob and an exotic Mexican herb called *pipicha,* and he details the process of making it so thoroughly that even a novice could pull it off. For seasoned bread bakers the going will be more intuitive, but either way, the results will be revelatory. I cannot wait to try it as a base for a BLT. Or a dunk for chili. Or simply thickly sliced and swiped with softened butter.

If there is a spectrum of cooking knowledge and preparation time needed for the dishes in this book, Tom Murphy's succinct "recipe" above and Grant Johnston's near-treatise on bread lie on either end of it. Between them are some 125 recipes geared mainly to people with average cooking skills, contemporary time limitations and a penchant for real-deal fare—as in made from scratch, not too fancy and deeply delicious. Meri Tunison's version of *domada,* a chicken, vegetable and ground peanut stew that is the national dish of Gambia immediately comes to mind. So do Philip Kauth's Brazilian-style garlic collards, Laura Schmidli's asparagus with potatoes, cherry tomatoes and Nepali spices, and Betty Rosengren's *bara brith,* a currant-studded Welsh tea bread.

If you need a term for the overall kind of cooking in this book, try "pan-populist" on for size. Featuring dishes like easy-going salads, multi-layered stews, lively side dishes and homey desserts, this is fare made with regionally sourced ingredients, brightened and broadened by a global outlook. It's cooking that is of the people, by the people, for the people.

Re-Visions

Recipe editing is a tricky business. Think of a dish—any dish, really—and then ask five people to write down how to make it. You'll end up with five surprisingly different-sounding formulas. Now multiply those writers by more than a hundred, each of them providing their best, often much-beloved recipes, and try wrangling all those takes on culinary instruction into one cohesive recipe style. My job with this book was to edit the recipes for clarity and consistency, so that cooks of every skill level can succeed at re-creating the dishes. But I also worked to retain the individual voices of contributors so that readers can experience the often entertaining or illuminating differences in their approaches.

You'll see standardized ingredient descriptions, measurements and basic cooking procedures throughout the recipes. For ingredients like hot chilies, garlic and such, there's frequently a range of amounts, to allow for personal preference. Here and there I added substitutions that favor market ingredients (say, honey instead of sugar, or dried cranberries instead of raisins), keeping true to the essence of the dish but giving choices where they might be welcome. Likewise, in the recipe directions, I offered alternative methods or types of kitchen equipment to give cooks more leeway—for example, using a hand-held potato masher instead of a food processor to puree pumpkin flesh.

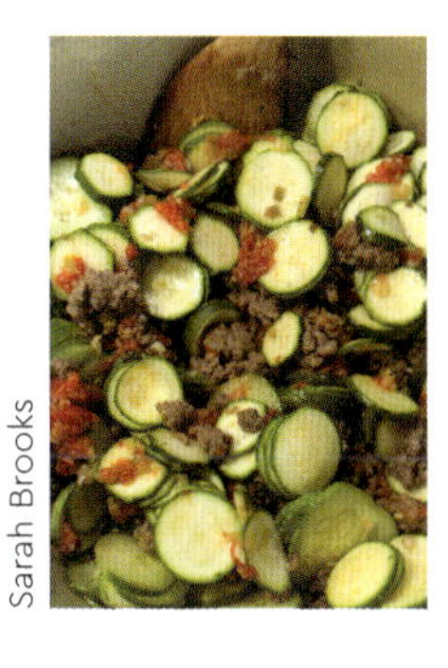
Sarah Brooks

...some 125 recipes geared mainly to people with average cooking skills, contemporary time limitations and a penchant for real-deal fare—as in made from scratch, not too fancy and deeply delicious.

Fair is fair, of course, meaning you get to do some of the same, if you like. There's nothing wrong with tweaking a recipe; in fact, that's exactly what good cooks do. If you're new to a dish—or to cooking itself—it's always wise to follow the recipe closely the first time around. *New York Times* food editor Sam Sifton says that this is like playing a cover song. But cook it again, or a few times, and "you're playing a new arrangement, even if it's relatively faithful to the original." Go ahead and replace the green beans called for with the asparagus you have on hand. Up the amount of serranos for the heat-seeking guests you've got coming over, or thicken a tomato sauce intended for pasta and use it on pizza instead.

The main goals here were to give you enough detail to prepare a dish successfully—the way the contributor intended—and to cheer you on if you choose to modify, substitute, play with or otherwise make the dish your own.

While all the recipes feature ingredients available (at least part of the year) at the DCFM, most of those foodstuffs are not exclusive to the DCFM. You certainly can use sauerkraut, or eggplants, or chili oil from other farmers markets or food sources. While this book celebrates the DCFM specifically, the recipes are for anyone who wants to enjoy local foods for any of the excellent reasons there are to do so.

Open Season

You already may have observed that the book's chapters aren't arranged around the seasons, or by ingredients. These formats steer most cookbooks that focus on farmers markets or local foods, and with good reason. Seasonality and local flavor are the mainstays of market-based cooking. Yet that approach has expanded, due in large part to the steady growth of value-added products at the market.

We formatted this cookbook somewhat traditionally—by meal course—with the implication that many recipes can be prepared throughout the year, or at least outside "harvest time." Frozen sweet corn, graham crackers, pastry flour, polenta, tomato sauce, kimchi, chutneys, focaccia, chorizo, dried curry leaves, tamales, salad dressings, soup stocks and so forth—these are some of the market goods that belie the strictly seasonal approach to cooking with local foods.

Besides frozen, canned, dried, fermented, packaged and otherwise lightly processed foods, the market boasts "off season" crops grown hydroponically or in cold-weather hoophouses. Throughout the year, ingredients like mushrooms, meats, poultry, eggs, baked goods, cheese, and on and on, are also available. Not everything at the market is offered every week, but the array is always extensive.

MISSION: POSSIBLE

The Dane County Farmers' Market was founded to pursue these goals:

- Give growers and producers of Wisconsin agricultural commodities and other farm-related products alternative marketing opportunities.
- Promote the sale of Wisconsin-grown farm products.
- Improve the variety, freshness, taste and nutritional value of produce available in the Madison area.
- Provide an opportunity for farmers and people from urban communities to deal directly with each other rather than through third parties, and to thereby get to know and learn from one another.
- Provide an educational forum for consumers to learn the uses and benefits of quality, locally grown or prepared food products.
- Provide educational opportunities for producers to test and refine their products and marketing skills.
- Enhance the quality of life in the greater Madison area by providing a community activity which fosters social gathering and interaction.
- Preserve Wisconsin's unique agricultural heritage and the historical role which farmers markets have played in it.

—From the Dane County Farmers' Market Website, DCFM.org

Regionally raised, peak-season produce will always be the mainstay of the market—its biggest draw and best asset. But this is just the beginning, really. The market season is every season.

Community Center

"We can talk about all the benefits of the market—social, environmental and economic—but in the end they all become personal. I know of no other large event that carries the sense of unity that this market seems to inspire."

—from *The Dane County Farmers' Market: A Personal History,*
by former DCFM manager Mary Carpenter

After more than fifty years of operation, the DCFM remains a market that requires its members to grow, raise and/or produce everything they sell. Crafts are not allowed, but non-edibles that are agricultural products are permitted, such as wool yarn, gourds and ornamental corn. The organization is also proud of its signature "producer-only" policy, meaning that farm and business owners must be at their stands on market days. The great value here is that the person who raised or made the food—who knows it best and is ultimately responsible for it—is there to engage with customers, answer their questions, and solve problems (or prevent them altogether). In turn, owners meet and learn from their customers. Ideas and advice are exchanged. Trust builds. Relationship happens.

In his book, *Taste: My Life Through Food,* actor-gastronome Stanley Tucci talks about the importance of buying food from small food-shop keepers like butchers and cheesemongers:

> To me, eating well is not just about what tastes good but about the connections that are made through the food itself. I am hardly saying anything new by stating that our links to what we eat have practically disappeared beneath sheets of plastic wrap. But what are also disappearing are the wonderful, vital human connections we're able to make when we buy something we love to eat from someone who loves to sell it, who bought it from someone who loves to grow, catch or raise it. ... Great comfort is found in these relationships, and they are very much a part of what solidifies a community.

It's the same at the farmers market, only more so, perhaps, because the relationship to food source is that much more direct, that much more meaningful. The market isn't just a place where food is bought and sold. It's a community of people who depend on each other—for livelihood or sustenance, yes, but also for conversation and networking, for sharing recipes and growing tips, for kinship.

When we asked contributors to share stories from the market, it was no surprise that many of their tales were about people connections. Natascha Merten told us about the day she returned to the market after being out of town for three weeks. "I literally got shouted at by three different vendor-friends [who] asked where I had been and how I could not [have] let them know. They had been seriously concerned," she wrote. "I will not do that again, but instead tell them now all the (unnecessary) details about my travels."

(I can sympathize with Natascha. I like to get to the market when it opens, when it's not too busy yet and there's more time to chat with vendors. But you should hear the ribbing I get if I arrive any later than 7 a.m. "Overslept, huh?" "Somebody was out too late again last night!")

Market regular Kristin Korevec's favorite market memory is from her wedding day (yes, she and her soon-to-be husband made time for the DCFM even on that momentous morning). "Larry [Haas], the 'Gourd Guy,' presented us with a giant knotted gourd signifying our marriage," Kristin relayed. This reminded me of another story I heard years ago from former meat vendors John and Dorothy Priske. This time, the gift-giving went the other way across the vendor table. Two faithful, early-morning customers, after having their usual chat with the Priskes, went home and baked a pie with seasonal apples they had just purchased at the market. Then they packed it up, still warm, drove back to the market and presented it to John and Dorothy.

Lucinda Ranney

Gift-giving, remembering birthdays, pitching in, sharing rituals, even practical jokes—these are what create bonds at the farmers market. Linked to what we cook and eat, they are ties that nurture community in a particularly elemental way.

This kind of care for each other is not unusual. I've seen customers purchase donuts and coffee for vendors too busy to leave their stand, and shoppers who stand in for a seller while the latter takes a break. I've watched growers weigh up pears, tell the customer the amount due and then add a couple more pieces of fruit to the bag.

Shopper Guy Downs Plunkett III wrote that his wife, Joanne, prefers apple crisp over cake for her late-July birthdays. Their "go-to apple guy" is Bob Willard of Ela Orchard, and they look to him for Lodis to use in her birthday dessert each year because that is the first apple varietal of the season. "Every year, Bob asks me if the Lodis came in on time [for Joanne's birthday]," said Guy. "I don't expect him to remember the date, but he remembers that the Lodis and her birthday come at about the same time."

Sometimes market camaraderie becomes habitual, even ritualized. Retired Dane County employee Christine Ladell passed along how she and other workers from a downtown government building used to gather regularly at the nearby Wednesday market, searching for a little sunshine break and some fresh ingredients for a shared lunch. "The rules provided we'd walk twice around the market perusing the best ingredients and then buy on the second go-around," Christine told us. "Then we'd meet in the lunchroom and individually 'design' our grilled cheese, awaiting our turn to use the sole grill. We would brag about the cross-hatch design we were able to get on our sandwiches. Many a county employee would walk down the first floor wondering from where that fabulous smell emanated."

One Wednesday, when interns from China were in town, the market lunch gang included the visitors in their grilled cheese sandwich routine. "They weren't accustomed to eating as many dairy products as us," Christine noted. "They couldn't believe how much cheese we ate." Most of the office lunch crowd are retired now, but I keep imagining that the aroma of molten cheese is still hovering in the hallways of county government.

Gift-giving, remembering birthdays, pitching in, sharing rituals, even practical jokes—these are what create bonds at the farmers market. Linked to what we cook and eat, they are ties that nurture community in a particularly elemental way.

Stanley Tucci, in his book quoted above, reminded me about the term "third place." It was coined by sociologist Ray Oldenburg to label public gathering places such as restaurants and bars, places that Oldenburg argues are as central to human well-being as home and workplace. I believe that the Dane County Farmers' Market is a vitally important third place in our region.

Furthermore, I take heart in knowing that there now are going on 9,000 farmers markets operating around the country. That's thousands of third places where people can access high-quality, healthful food, bolster regional agriculture and economies, and just plain have a good time together. These community-strengthening settings are especially significant in the years following a pandemic, when so many third places were taken from us.

Good food builds bridges—from farm to table, from market to community and, as this book illustrates, from recipe to world. When we connect the dots, we recognize not only that everyone eats, but everyone sits at the same table, too. Everyone belongs.

Welcome to this celebration of the Dane County Farmers' Market community: founders, staff, board members, growers, bakers, makers, gleaners, shoppers, cleaner-uppers, chefs, home cooks, volunteers and past vendors. Let's salute the market's enormous diversity of regional foodstuffs. Let's relish our differences with recipes from around the globe. And let's tip a hat, gratefully, to farmers markets everywhere.

Terese Allen
2023

Lucinda Ranney

Lucinda Ranney

Lucinda Ranney

Terese Allen

APPETIZERS AND DRINKS

STUFFED JALAPEÑOS WITH BACON AND CHEVRE

4 TO 6 SERVINGS

You know them as jalapeño poppers, but think of them as a north-of the-border snack version of Mexico's chiles rellenos. The list of ingredients for this perennial favorite isn't long, and you can get nearly all of them at the farmers market. This version came from goat cheesemaker and market vendor Felix Thalhammer of Capri Cheese, and we're passing it along to you with a reminder—jalapeños can vary tremendously in their level of heat. We recommend you don't pop a whole one in your mouth. Enjoy a nibble of it first, and then take it from there.

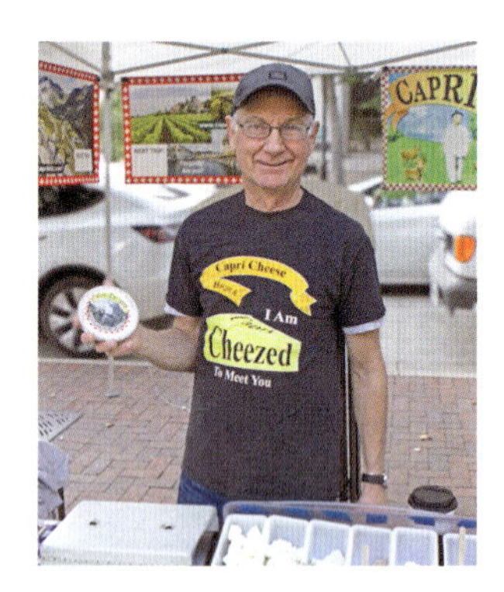

2 to 3 ounces (1/2 to 2/3 cup) fresh goat cheese (chevre), at room temperature

2 ounces (1/4 cup) cream cheese, at room temperature

1 tablespoon minced shallot or 2 tablespoons minced green onion

1 teaspoon minced garlic, sprinkled with a pinch of salt and pressed to a paste with a fork or the back of a knife

1/4 teaspoon each salt and pepper

8 medium jalapeños, cut in half lengthwise and seeded

2 to 3 slices bacon, cooked, drained on paper towels, cooled and crumbled

2 tablespoons hot pepper jelly or jam

Heat broiler to high. Line a baking sheet with aluminum foil. To make the filling, place both cheeses, shallots or green onions, garlic, salt and pepper in a bowl. Stir until well-combined.

Arrange jalapeño halves on the baking sheet and portion the filling into the peppers. Broil until peppers are beginning to brown and are tender, 5 to 7 minutes. Meanwhile, whisk jelly and 1 teaspoon water in a small cup; microwave until jelly is melted, about 20 seconds.

To serve, transfer peppers to a platter. Sprinkle with crumbled bacon and drizzle with jelly or jam. Serve hot.

Rachel Figueroa

Lona Alsum

ALSUM'S CHEESY CORN DIP

10 OR MORE SERVINGS (ABOUT 3 1/2 CUPS DIP)

Hearty stews and slow-simmered soups are fine in winter, but during those long, often dull-tasting weeks, do you ever find yourself craving something sun-kissed and summery? Yeah, we thought so. Alsum Sweet Corn to the rescue here. It's a fourth-generation family farm that offers fresh corn in season and frozen corn year-round at the market.

This recipe came to be one afternoon when Lona Alsum and her daughter Brittany did some experimenting to create a special dip for a family gathering. Now it's one of their clan's most requested recipes and a fond memory of Alsum creativeness.

"By adding different spices from almost any ethnicity to our basic recipe, you can create your own unique blend and taste," says Lona. Think: Garam masala or curry powder; cumin and fresh cilantro; Cajun seasoning; dill and caraway. To hoist the heat, replace the plain frozen corn called for in the recipe with their "Sweet Heat Corn," which has jalapeños in it.

1 package (8 ounces) cream cheese

1 bag (16 ounces) frozen sweet corn, thawed in the bag with its juices

1/3 cup sour cream

1/2 cup shredded cheddar

1/4 cup coarsely grated Parmesan

1 tablespoon dried oregano

1/2 teaspoon black pepper

1/2 teaspoon dried chile flakes (optional)

1/2 teaspoon garlic powder (optional)

Tortilla chips

"...you can create your own unique blend and taste."

Place cream cheese in a saucepan over medium-low flame and let it heat up, stirring occasionally, until it's almost melted, 5 to 6 minutes. Add corn with its juices, sour cream, cheddar and Parmesan. Continue to heat the mixture, stirring often, for another 2 to 4 minutes. When it's combined well, stir in oregano and black pepper, and the dried chile flakes and garlic powder, if using. Serve hot or warm with tortilla chips. Leftovers, if there are any, are easily reheated in the microwave (and they're even good cold).

YAKITORI SHISHITO PEPPERS

2 TO 4 SERVINGS

Ever since shishito peppers hit the market scene in the late 2010s, there's been no going back. Thin-skinned, nearly seedless and shaped like slender, crooked thumbs, with stems that double as built-in handles, shishitos make the perfect appetizer when blistered whole over high heat and served with a dipping sauce such as sriracha mayo, sesame sour cream or lemon aioli. Their heat level is low—usually, that is. Every once in a while, a few rogue spicy ones will flash-mob a batch of peppers, creating a good-natured game of Russian roulette for diners. This is fun if you like surprises; otherwise, proceed with caution.

The recipe here is from Gilbert Altschul, chef-owner of Grampa's Pizzeria, who frequently enjoyed shishitos at a *yakitori* restaurant in San Diego when he lived there. (Yakitori is Japanese skewered grilled chicken, typically eaten as street food or in casual restaurants.) In Madison, when he began finding shishitos at the DCFM, he recreated the memorable flavors with this recipe.

Gil serves the zesty bites on a wooden board and washes them down with Japanese lager. "A little drizzle of Thai chile fish sauce is great, too, for those who want to up the heat," he adds.

NOTE: Several ingredients called for here may sound unfamiliar. Look for them in Asian food shops and at larger grocery stores.

28 to 30 shishito peppers

About 2 teaspoons *shoyu* (Japanese soy sauce)

About 1 teaspoon *yuzu* sauce (or *ponzu*, a sweet-salty sauce made from citrusy yuzu fruit)

1/4 to 1/2 teaspoon grated lime zest

About 1/4 teaspoon *togarashi* (zingy, toasty Japanese seasoning mix)

1 to 2 teaspoons *katsuobushi* (dried bonito flakes)

Prepare very hot coals in a charcoal grill. Skewer shishitos onto long bamboo skewers.

Blister the peppers over the coals, turning them often as they char. This should take only a minute or two. While they're still hot, remove the peppers from the skewers and put them in a mixing bowl. Working quickly, drizzle in the shoyu, yuzu sauce, lime zest and togarashi, and toss well. Place peppers on a wooden board and sprinkle liberally with bonito flakes. If the peppers are still nice and hot the bonito will "dance" from the heat. Serve immediately.

Lucinda Ranney

Lucinda Ranney

MUHAMMARA
(ROASTED RED PEPPER & WALNUT DIP)

MAKES 2 TO 3 CUPS

"Be careful! This is quite addictive," warns DCFM shopper Betsy Abramson. Roasted red peppers, toasted walnuts, garlic and Middle Eastern seasonings combine in a rich, spicy dip. Use it with toasted pita chips, crudités or crackers, in omelets, on grilled meats or fish, or even as a pizza sauce.

Muhammara, which means "reddened" in Arabic, originated in Syria and Lebanon and can be found on many a mezze platter around the Middle East. Traditional recipes call for a little pomegranate molasses, but pure maple syrup is a lot easier to find in our corner of the world, and it works well, too.

Muhammara freezes nicely, Betsy says. "So grab four of those beautiful DCFM red peppers at the end of the season and double your batch."

"Be careful! This is quite addictive..."

2 large red bell peppers

1 1/2 cups raw walnuts

3 medium garlic cloves

2 teaspoons paprika or 1 teaspoon Aleppo pepper flakes

1 teaspoon ground cumin

1 teaspoon dried chile flakes (or less if you're not into spicy)

1/4 cup fresh lemon juice

2 tablespoons maple syrup or pomegranate molasses

2 tablespoons extra-virgin olive oil

1 teaspoon kosher salt

Freshly ground black pepper

Additional olive oil, optional

Heat oven to 400 degrees. Line a baking pan with foil or parchment paper, place the peppers on it and roast them, turning them every 10 minutes or so until the skin blackens and blisters. Place them in a bowl, cover with plastic wrap and let stand for at least a half hour. (This creates steam to loosen the skins.)

Meanwhile, decrease oven heat to 350 degrees. Line a baking sheet with parchment and spread the walnuts on it. Roast the nuts but watch them carefully so they don't burn. It should take 10 to 12 minutes at most. Remove from the pan and allow to cool.

Peel the red peppers with a paring knife or your fingers; scrape off and discard stem and seeds. Coarsely chop the peppers.

Switch on a food processor and drop the walnuts and garlic through the feed tube. Then toss in everything else, including the red peppers, but not the additional olive oil. Keep it running until the mixture is smooth, pausing once or twice to scrape down the sides of the work bowl. Taste and adjust seasonings as you like.

Transfer mixture to a bowl, cover it and, if time permits, let it chill in the fridge for a couple hours to develop flavor. Return it to room temperature before serving. To serve, drizzle a little more olive oil over the surface, if desired.

Lois Bergerson

AIR-FRIED OYSTER MUSHROOMS

4 TO 6 SERVINGS

Looking for an upscale appetizer for a special meal? Here it is, thanks to longtime DCFM vendor Jamie Ramsay of Indian Farm Mushrooms, who in turn thanks Madison chefs Ben and Jonny Hunter for creating it.

"This is our favorite appetizer from the [Hunters'] former Forequarter restaurant in Madison," says Jamie. It features Indian Farm's delicate oyster mushrooms (yellow, gray, brown or pink) dipped in an Asiatic shallot sauce and air-fried, then accented with a black garlic puree. Yowza. The finished mushrooms can also be added to Asian-style noodle or rice dishes and served as a main course.

Cooking notes:

The recipe calls for an air-fryer, but if you don't own one of these countertop appliances, you can also use the air-fryer setting in a stove oven. With the air-fryer setting turned on, heat the oven to 400 degrees, arrange all the mushrooms on a rack over a sheet pan, and air-fry them about 15 minutes.

You'll have leftover sauce after the mushrooms have been fried. Use it for another batch of mushrooms or other air-fried veggies, or as a marinade for grilled chicken breasts or steaks.

Terese Allen

For the optional black garlic puree: Black (or aged) garlic can be found in specialty food shops and in the gourmet aisles of some grocery stores, as well as at the Hickory Hill Farms stand at the DCFM. To make a black garlic puree, peel black garlic cloves and soak them in hot water. Drain off the water into a bowl. Puree the garlic in a blender or food processor with just enough of the reserved water to make the mixture smooth. Alternatively, use a bottled black garlic puree instead of making your own.

1 pound oyster mushrooms

1 recipe (about 1 1/2 cups) Caramelized Shallots (see sidebar on page 30)

1 1/2 cups plus 1 tablespoon water

1/3 cup bottled fish sauce

1/4 cup rice vinegar

2 1/2 teaspoons sugar

2/3 cup canola oil

Bottled or homemade black garlic puree (optional)

Cut mushrooms from the stalk and clean them gently with a damp paper towel.

Make the sauce: Place caramelized shallots, water, fish sauce, rice vinegar and sugar in a blender or food processor. Blend at high speed until pureed, then *slowly* add the oil in a thin stream. The mixture will be rather thick. Transfer it to a large bowl; use a rubber spatula to fold mushrooms into the mixture and coat them well.

Fry the mushrooms: Preheat an air fryer to 400 degrees. Heat stove oven to 250 degrees (it will be used to keep the first batch of mushrooms warm). When air fryer is ready, place half of the mushrooms—selecting ones of similar size—on the fryer tray. Fry them for 10 minutes, then check to determine if they're done to your liking. Increase the frying time if needed. It may take as long as 15 minutes for larger mushrooms to be done. The cooked mushrooms should be well-browned and crisp; take care not to burn them. When they're done, remove them to a rack set over a sheet pan; keep them warm in the oven while you fry the second batch.

While mushrooms are cooking, smear a teaspoon or two of black garlic puree, if using, on the side of six appetizer plates. When all the mushrooms are done, divide them onto the plates and serve immediately.

"Use leftover sauce as a marinade for grilled chicken breasts."

CARAMELIZED SHALLOTS

MAKES 1 1/2 TO 1 2/3 CUPS

One of the tricks that chefs employ to make savory dishes more delicious is to use shallots in them. Shallots are sweeter than yellow onions, and when caramelized, they can make a good dish great. They're a main feature in the recipe for Air-Fried Oyster Mushrooms (see page 28) and will boost flavor in everything from omelets and grilled sandwiches to pastas and vegetable tarts. Look for shallots at market from late summer into winter (they store well) and don't just buy a couple—buy a pound or three to step up your meals for weeks to come.

2 tablespoons unsalted butter

1 tablespoon olive oil

1 1/2 pounds large shallots, peeled and sliced

Salt

Place a large skillet over medium-high flame; add butter and olive oil. When butter begins to sizzle, add shallots and a couple pinches of salt. Stir well to coat shallots with fat. Cook, stirring occasionally, while shallots give off liquid, 10 to 15 minutes.

Once most of the moisture is gone, they will start to brown and smell heavenly. Reduce heat to medium and continue to cook another 15 to 25 minutes, stirring often as they brown, adjusting the heat so they don't burn, and scraping up the darker, flavor-packed caramelized areas—called fond—that form on the bottom of the pan. If things get too dry, you can add a teaspoon or two of water. Doneness is your call: Caramelized shallots can be golden brown, medium brown, or deep brown, with the latter being the most deeply flavored.

QUICK-PICKLED SHALLOTS AND RADISHES

MAKES ABOUT 2 CUPS

These tangy nibbles are adapted from an America's Test Kitchen recipe. "I use them on salads, tacos and sandwiches. It seems to go well with Mexican foods but also has Asian possibilities," says contributor Sue Allen. "I like having something low-salt and tangy to take the place of traditional American pickles, which I love. But something different is nice."

Her take on the original recipe is to mix purple daikon radishes with red ones. "The coloring is beautiful," she says. Sometimes she uses turnips and kohlrabi. "Even with these white veggies, the shallot turns things a lovely shade of pink after it sits awhile." You can eat the pickles soon after making them, or let them sit in the fridge overnight to let the veggies soften and become infused with the lime juice.

Sue lives an hour from Madison and has a large vegetable garden. "But the DCFM is still my inspiration, and I try to get there at least once or twice a season for the fun of it."

Lucinda Ranney

"...the DCFM is still my inspiration and I try to get there at least once or twice a season for the fun of it."

1 to 2 cups thinly sliced red radishes and purple daikons

1/4 cup thinly sliced shallots

Juice of 2 limes

1 teaspoon sugar

1/8 teaspoon salt or to taste

Combine all ingredients in a bowl. Serve right away, in 10 minutes, or a half-hour, or refrigerate them and eat them the next day. They will keep well longer than that, but they may not last that long.

LIQUID SUNSHINE
(AKA GINGER TURMERIC TEA)

MAKES ANY NUMBER OF SERVINGS

Recipe contributor Sarah Elliott is a former DCFM manager who is now a regular customer at the market. Says Sarah: "While I now enjoy my Saturdays as a shopper, the time I spent as manager is something I hold dear. The camaraderie and fellowship that comes with being at the market every single week—rain, shine, cold, snow, sunny, dark—creates a very special bond."

When fresh ginger and turmeric roots first started making an appearance at market, Sarah couldn't resist them. "I had to buy a handful of the knobby delicacies. Both are delightful in curries and stir fries." Sometimes Sarah has extra knobs that don't make it into a cooked dish, so she juices them and freezes the juice in ice cube trays. Later, she dissolves the golden cubes in hot water for a vibrant tea that "helps carry the warmth of summer through the winter months."

Approximately equal parts of fresh ginger and turmeric roots

1 unpeeled apple (needed only if you're using a juicer machine), seeds removed

Local honey

Thoroughly scrub ginger and turmeric roots to remove any debris.
There is no need to peel them.

Sarah Elliott

Sarah Elliott

If you have a juicer, run the ginger and turmeric through the juicer, followed by the apple (to help make sure you include all the good bits). Pour juice mixture into an ice cube tray and freeze.

If you don't have a juicer, thinly slice the ginger and turmeric roots and add them to a pot of water—approximately 1 1/2 inches each of ginger and turmeric for every cup of water. (There's no need for the apple with this method.) Bring to a simmer, partially cover and cook low and slow for 20 to 30 minutes. Strain mixture through a fine-mesh sieve over a bowl, then pour the liquid into ice cube trays and freeze.

Once the cubes are frozen, place them in a single layer in freezer bags for easy storage. (And be prepared for your ice cube tray to turn a little yellow from the turmeric. Promptly removing the cubes once they're frozen and soaking the tray in a little vinegar helps; however, they may remain a little stained.)

To serve: In the depths of winter or any time you need a little sunshine in your day, put a cube of the frozen juice into a mug, fill with hot water (no need for it to be boiling) and add local honey to taste. Relish a vibrant, warming, locally-grown treat!

FREEZE THAT FRESH GINGER

"I think I used to keep ginger in the fridge, but I could not tell you when. I have long since swapped refrigerated ginger for frozen. Not only does this prevent sad, shriveled and moldy gingerroot hidden in the bottom of your produce drawer, it also makes for easier prep and more even flavor distribution. When I buy a new knob of gingerroot, I wash it, break it into pieces and stick it in the freezer. Then, whenever I want to use it, I just shave some off with a Microplane [grater]. No strings and no peeling necessary. And if you want to have sliced fresh ginger, just let some thaw for a bit and then slice or chop as much as you need."

—DCFM shopper Megan Bjella

THIS IS A SPICY DRINK

MAKES 1 COCKTAIL, AND CAN BE MULTIPLIED TO SERVE MORE

Here's a rousing example of how the farmers market can inspire a cook. It also may be the "chef-iest" recipe in this book—an inside view of professional brainstorming and the culinary resources a chef draws from to execute an idea. Its creator, Gilbert Altschul, buys from the market for his restaurants (Grampa's Pizzeria, Bandit Tacos and Coffee, and Gib's, a cocktail bar, all in Madison), working with growers to learn what crops grow best on their land, the ones that taste best.

"I really fell in love with the lemon drop peppers [from Savory Accents] and decided to build a cocktail around it," Gil says. "I love spice and wanted a truly spicy drink."

He nailed it. The lemon drop is a small, yellow-when-ripe varietal with a citrusy tang and significant kick. Gil macerates the peppers in good vodka for a week or so to create a peppery base for the drink. He juices in-season cucumbers to use as a cooling counterpoint and adds fresh lime juice for thirst-quenching tartness. Then he builds in the rest: *Pisco*, a clear brandy from Peru; pineapple gum syrup (or gomme), which gives the drink a silky texture and tropical sweetness; and curaçao, a specialty liqueur made from the peels of bittersweet oranges (Gil recommends Pierre Ferrand Dry Curacao).

Gil gave the drink a name he says is "very on the nose, so that folks couldn't say that they weren't warned, if it was indeed too spicy for them." But we doubt anyone will complain. It's a knock-out craft cocktail, as appealing as it is startling in concept, appearance and flavor.

1 ounce (2 tablespoons) lemon drop chile vodka (see method on page 36)

1 ounce (2 tablespoons) pisco

Scant 1 ounce (scant 2 tablespoons) pineapple gum syrup
(purchase online or at specialty stores)

3/4 ounce (1 1/2 tablespoons) cucumber juice (see method on page 36)

1/2 ounce (1 tablespoon) fresh lime juice

Scant 1/4 ounce (scant 1 1/2 teaspoons) curaçao

Combine all ingredients in a metal cocktail shaker; fill with ice and shake. Double strain it over fresh ice into a tumbler or Old Fashioned glass and serve immediately. (To double strain the cocktail, place a cocktail strainer on the cocktail shaker and hold those together with one hand. Then with your other hand hold a fine mesh strainer by its handle over the serving glass. Pour cocktail through both strainers into the glass.)

"I really fell in love with the lemon drop peppers..."

Erika Whitson

TRANSLATING CHEF-SPEAK

The cocktail recipe from Gil Altschul on page 34 features ingredients that he makes from scratch, such as chile vodka and cucumber juice. When we asked Gil for these "recipes within the recipe," the instructions he sent were concise ... but maybe not so do-able for home cooks without the special equipment and skill set that chefs possess. So here's how to read a chef's mind and cook at home like a pro:

How to Make Lemon Drop Chile Vodka

Chef Gil says: "Vacuum seal 6 sliced lemon drop chiles from Savory Accents with 750 vodka (we use Tito's) and cook in circulator for 3 hours."

Translation (for home cooks who don't own a circulator): "Slice up 6 lemon drop peppers (available August through October from Savory Accents at the Dane County Farmers' Market). Combine the peppers—seeds and membranes included—with 1 bottle (750 milliliters) Tito's Vodka in a clean one-quart Mason jar. Attach the lid tightly and leave to stand at room temperature until desired level of spiciness is achieved, 5 to 7 days. Line a sieve with cheesecloth, set it over a bowl and pour the vodka through it to strain out the peppers. Transfer vodka to a clean bottle with a tight lid. It will keep for months."

How to Juice Cucumbers

Chef Gil says: "Cucumbers can be juiced in a masticating or centrifugal juicer."

Translation (for folks who don't have one of those, either): "Scrub a medium-large, 3/4-pound cucumber, cut off any blemishes on the skin and coarsely chop it up. Place the pieces in a blender or food processor and puree at high speed until as liquified as possible. This could take up to a couple of minutes, and if the cuke isn't juicy enough to create a slushy liquid, you may need to add a small amount of water. Line a sieve with double cheesecloth that is big enough to hang over the sides, and set the sieve over a bowl. Working in batches if necessary, pour in the cucumber slush, gather up the overhanging cheesecloth, then twist and squeeze to extract as much juice as possible. Extra juice will keep a few days in the refrigerator, or it may be frozen (or iced and enjoyed on its own). Note: If you don't have a blender or food processor, you can grate the cukes on a hand-held grater, then proceed as above. A 3/4-pound cucumber should yield about 1 1/4 cups juice.

How to Make Pineapple Gum Syrup (or Gomme)

Chef Gil says: "Pineapple gomme is very laborious. It may be simplest to purchase it [online at Liber & Co.]"

Translation: "What he said."

SOUPS

Recipe on page 56

Terese Allen

CELERY AND BLUE CHEESE SOUP

4 LARGE OR 5 SMALL SERVINGS

Stilton, named after a village in Cambridgeshire County, England, is a soft, crumbly, English blue cheese and the inspiration for this simple but elegant soup. Contributor Ally Shepherd features Tilston Point Blue Cheese from Hook's Cheese in it to echo the distinctively pungent flavor of Stilton. "This is my mum's favorite soup recipe and I always ask her to make it when I go back to the U.K. to visit," writes Ally. "Soups are such a great way to keep warm during British winters—and, therefore, during Wisconsin winters, too!"

You can use any of Hook's celebrated blue cheeses (see page 48)—or those from another company—as a substitute for the Tilston Point, and broccoli makes a nice alternative to celery, says Ally. To gild the lily, add chopped fresh chives or tarragon to the soup as it finishes simmering, or top each bowlful with them. Gilded or not, this is of those tasty comfort soups you might pair with a nice piece of chunky brown bread, maybe some apple chutney and a glass of stout or porter.

4 tablespoons butter

7 cups thinly sliced celery stalks (from about 2 bunches)

4 cups finely chopped onions (about 1 pound total)

1 tablespoon flour

2 cups chicken or vegetable stock

Salt and pepper

3 to 4 ounces Hook's Tilston Point or other blue cheese, crumbled into small chunks

"Soups are such a great way to keep warm during British winters—and, therefore, during Wisconsin winters, too!"

Heat butter in a soup pot over medium flame. When the butter is hot, add the onions and saute them until golden, stirring often, about 6 to 8 minutes. Stir in the celery, partially cover the pot and let the veggies sweat until they're soft, another 10 minutes or so, adjusting the heat lower if necessary to prevent browning.

Scatter 1 tablespoon of flour over the vegetables and stir for a minute over the heat. Gradually stir in the stock, along with 2 cups of water. Season lightly with salt and pepper and bring the soup to a simmer, then reduce the heat and simmer slowly for 30 to 40 minutes. Use a hand-held immersion blender, a food processor or a countertop blender to puree the soup. At this point, if you'd like a very smooth soup, pass it through a fine mesh strainer, then return it to the pot.

Reduce the heat to its lowest point and stir in the cheese a little at a time until melted. Taste the soup and add more seasonings, if needed. The flavor will develop nicely if you let the soup sit off the heat for about a half hour or so. Then gently reheat it just before serving. Do not let it boil, as the cheese will curdle.

Shutterstock

RED KABOCHA SQUASH SOUP WITH GINGER, LEMONGRASS AND THAI EGGPLANT

4 TO 6 SERVINGS

"I look forward to seeing red kabocha, lemongrass, Thai eggplant and fresh ginger appear on market stalls when the air is crisp, and a thick, hot soup would make a perfect lunch or supper," says James Wehn, a market-o-phile who lives near the Capitol Square. He revised this vibrant soup from a recipe his mother shared with him, creating one that's infused with flavors common in numerous Asian cuisines. "Except for the coconut milk and fish sauce, all the ingredients are available in early autumn at the market (including the stock, if you make it yourself)."

While other types of hard-fleshed winter squash can be used here, James likes red kabocha because the skin is often tender and edible and can be included right in the soup. If you want to make this during the cold months, just plan ahead: "Red kabocha squash can be roasted, cut up, and stored frozen," says Wehn. "The lemongrass stalks and Thai eggplant can also be frozen. Cut the Thai eggplant into pieces and freeze them. Do not thaw (which turns them mushy and brown), but simply put the frozen pieces directly in the simmering soup."

To round out the meal, do what James sometimes does—add the soup to a bowl with quinoa and a soy and ginger-braised chicken thigh.

Cooking note: If any of your diners are among the poor souls with DNA that makes cilantro taste like harsh soap, use Thai basil instead.

1 medium red kabocha squash (2 to 3 pounds) or substitute other orange-fleshed winter squash like butternut or honeynut

Olive oil for roasting and sauteing

1 to 1 1/2 cups chopped onions

1 tablespoon maple syrup or brown sugar

2 tablespoons grated fresh ginger

1 heaping tablespoon minced garlic

1 to 2 chile peppers (fresh or dried), seeded and finely chopped

3 to 4 stalks lemongrass (each about 6 inches long), lightly smashed with the flat of a knife, to release flavor

1 can (14 ounces) coconut milk

2 cups vegetable or chicken stock, or more as needed

1 teaspoon bottled fish sauce, or more to taste

5 to 6 Thai eggplants (about 10 ounces total)

Salt and freshly ground pepper to taste

1/2 cup chopped cilantro, divided

Heat oven to 400 degrees. Use a sharp, heavy knife to cut the squash in half; scrape out the seeds with a spoon. Lightly oil the flesh with olive oil and place the pieces cut side down on a baking sheet (lined with parchment paper, if you like). Roast the squash, turning the pieces over about 20 minutes into the baking time, until they're tender, about 30 minutes total. Allow squash to cool and cut the large chunks into bite-size pieces.

Heat 1 to 2 tablespoons olive oil in a soup pot over medium heat. Add the onions and maple syrup or brown sugar, and cook until onions are softened, 6 to 8 minutes. Stir in ginger, garlic, chiles and lemongrass, and continue sauteing 1 to 2 minutes to coax out their aromas and flavors.

Add roasted squash pieces, coconut milk, stock and fish sauce. Cut Thai eggplants into bite-sized pieces (you want about 2 to 2 1/2 cups) and add the pieces to the soup. Raise heat to medium-high, bring soup to a low boil, then reduce heat to maintain a simmer. Cook the soup, stirring regularly, until the eggplant is barely tender, 10 to 15 minutes. Add more stock (or water) as needed to adjust the thickness of the soup to desired consistency. Stir in salt and pepper to taste.

To finish the soup, remove the lemongrass stalks and stir in most of the chopped cilantro. Ladle soup into bowls and garnish with remaining cilantro.

James Wehn

James Wehn

CURRIED PUNKIN SOUP

4 TO 6 SERVINGS

Most people would use canned pumpkin puree here, but not farmers market enthusiasts. They want first-rate flavor, the kind that comes from fresh pie pumpkins, which are cultivars bred for flavor, not girth. Seek out varieties like small, round Early Sweet Sugar Pie pumpkins, elongated Nantucket Long Pies, or bumpy-skinned Galeux d'Eysines (also known as peanut pumpkins). To learn how to make your own pumpkin puree, see page 43.

This is a fairly spicy soup—not killer spicy, but it will wake you up a little. It comes from a collection of Trinidadian recipes that contributor Leslie Ann Busby-Amegashie has compiled.

- 4 tablespoons butter
- 1 cup chopped onion
- 1 to 2 teaspoons minced garlic
- 3 cups chicken broth
- 1/2 cup uncooked rice
- 1 tablespoon curry powder
- 1/2 teaspoon ground coriander
- 1/2 teaspoon cayenne pepper
- 1 3/4 to 2 cups cooked, mashed pie pumpkin or canned pumpkin
- 1 cup heavy cream
- Salt and pepper

Melt the butter in a medium-sized pot over medium heat. Add onions and garlic; cook, stirring often, until tender, 5 to 7 minutes. Add broth, rice, curry powder, coriander and cayenne. Bring to simmer and cook 20 minutes. Stir in pumpkin and cream; return to a low simmer and cook an additional 10 minutes. Season to taste with salt and pepper.

PUREE THAT PUMPKIN— AND OTHER WINTER SQUASH (ROAST THE SEEDS, TOO)

Making your own pumpkin puree is instinctual and user-friendly. First, pretend you're about to bake potatoes: Poke the squash(es) in several places with a meat fork and bake at 350 degrees until fully tender, about 1 1/2 to 2 hours, depending on size. Split them open to drain, let them get cool enough to handle, and then pull off the skins and scoop out the seeds and fibrous material. (Don't toss the seeds—see below for how to roast them for a snack.) Puree the flesh with a food processor or potato masher. Cool the puree completely and then freeze it in sealed containers.

(This method, of course, also works for most any kind of winter squash; the bigger they are, however, the longer they will take to bake fully. If you want roast-ier flavor and denser texture—and don't mind the extra work—you can cut open, seed and chunk up the squash before baking and pureeing it.)

With pumpkin puree on hand, you're all set for Thanksgiving pie, plus it can be used in soups, stews, smoothies, pancakes, risottos, even mac and cheese. Or just add butter and maple syrup and you've got a side dish. Pumpkin's sweet blandness takes to assertive complements like curry blends and chiles, but it also duos nicely with quieter flavors: nuts, cream, small amounts of sage or nutmeg.

Terese Allen

And about those seeds: When seasoned and roasted, the seeds from a baked—or raw—squash make a nutritious snack or a tasty garnish for soups and salads. Place the seeds in a colander under running water, swishing them around and pulling out the fibers. You won't get it all—just get as much as you can.

Spread them out to air-dry or dry with paper towels. Toss seeds with olive oil and sea salt (a little cayenne, too, if you like). Then spread them out on a baking sheet lined with parchment paper and roast in a 375-degree oven until golden brown, about 10 minutes, tossing once or twice as they bake. You can also soak the seeds in salted water overnight before drying off and roasting them. These won't need additional salt, as they will already be permeated with a light saltiness.

ISABEL'S COLLAGENIC BEEF SOUP

5 SERVINGS

Beef and pork vendor Matt Walter shared his wife Isabel's recipe for a deceptively simple beef soup. High in collagen from the bone marrow in it, the broth feels silky and subtly rich. The main flavorings—garlic, cilantro, paprika and cumin—are common in the cuisine of Isabel's home country, Honduras. She uses water and bouillon to create a light broth, but chicken or beef stock would also be welcome.

"Isabel developed this soup because it's low-carb and she believes in the health benefits of collagen," says Matt. "Isabel loves to cook, and we eat well every day!"

1 large, meaty beef soup bone cut from the shank (about 1 1/2 to 2 pounds), rinsed

5 to 6 cups unsalted homemade chicken or beef stock or water

2 teaspoons chicken bouillon granules or 2 chicken bouillon cubes (include this only if water is the liquid you're using)

1 medium onion, peeled, quartered and rinsed

5 medium or 3 large garlic cloves, peeled and rinsed

2 to 4 cilantro stalks (stems and leaves), rinsed

1/3 cup white vinegar

2 teaspoons salt

1 1/2 to 2 teaspoons ground pepper

1 teaspoon smoked paprika

1 teaspoon ground cumin

Garnish (optional): Chopped green onions and/or cilantro

High in collagen from the bone marrow in it, the broth feels silky and subtly rich.

Matt Walter

Combine soup bone and stock or water in a medium pot (if you're using water, add the bouillon, too). Add the onion, garlic, cilantro and vinegar. If the liquid doesn't cover the bone, add more stock or water so that it covers the bone by 1/2 inch. Bring liquid to a low boil over medium-high heat, skimming the surface as needed to remove the froth that forms.

Once the froth is no longer forming, stir in the salt, pepper, smoked paprika and cumin. Adjust heat so that the liquid boils slowly (you don't want a hard or fast boil here, just a nice lazy one). Partially cover the pot and cook soup about 3 hours, checking it about every 20 minutes and adding stock or water to keep the bone covered in liquid. As it cooks, the heat coaxes out beef flavor and healthful collagen from the bone.

After about 3 hours, the meat will be tender. Remove the bone and let it cool on a plate. Strain soup liquid through a cheesecloth-lined strainer into a clean pot; discard the solids. Return soup to a simmer. Taste and adjust the seasonings—it should have a mild tang from the vinegar and a little heat from the paprika, and should be on the peppery side. Taste for salt, too, and add a little more, if necessary, to heighten the flavors.

Remove meat from the bones, pulling off any cartilage or fatty scraps, and shred the meat with a fork or chop it with a knife. Or leave it in larger pieces if you like. Use the end of a spoon to pull the marrow out of the bone; chop it up and add it to the soup along with the meat.

Serve the soup piping hot. Chopped green onions and/or cilantro would be delicious on top.

ZEAMA FROM MOLDOVA (CHICKEN VEGETABLE SOUP WITH NOODLES AND FRESH HERBS)

8 TO 10 SERVINGS

Here is a traditional Moldovan chicken noodle soup, one that's served with the chicken *on* the bone in the bowl. It's typically made with *borş*, a fermented wheat bran juice used in Romanian and Moldovan cuisines, but lemon juice is a good substitute to give the soup its characteristic lightly sour taste. Another defining flavor is lovage, the intense celery-tasting herb found occasionally at farmers markets. (Or use celery leaves instead.) Vegetables, thin noodles and additional herbs give it more savor. The customary garnishes are sour cream and spicy red pepper, and cornbread is usually served on the side.

Market regular Irina Punguil Bravo, born and raised in Moldova and now a Madisonian, is the contributor. "*Zeama* is often made after a wedding celebration," says Irina. "Family and friends come to the bride's house to eat this delicious soup, as it's the perfect hangover cure. After a bowl of zeama, the wedding celebrations continue."

Zeama has been a constant throughout Irina's life. "This recipe is my nostalgic taste of summers spent at my grandma's in the village, and my mom's cure for colds. When I moved [away] to study in Romania, it was my food request every time I came back to Moldova on my breaks. Now it is the recipe I make for my daughter regularly."

1 small (about 3 1/2 pounds) organic chicken

1 tablespoon salt, plus more as needed

3 dried bay leaves

1 1/2 cups diced tomatoes

1 1/2 cups diced red or green bell peppers

1 1/4 cups diced celery

1 cup diced carrots

1 cup diced onion

2 tablespoons minced garlic

1/4 cup finely chopped fresh dill

1/4 cup finely chopped fresh parsley

2 tablespoons finely chopped fresh lovage (or 1 tablespoon dried) or 1/4 cup finely chopped celery leaves

2 cups fine egg noodles or thin spaghetti broken into thirds

Juice of 1 medium-large lemon

Garnishes (optional): Sour cream and dried chile flakes

Use a meat cleaver or kitchen shears to cut the chicken into medium-sized, bone-in pieces (about 12 total). Rinse the pieces, place them in a large soup pot and cover them well with water (about 12 cups).

Bring the water to a low boil over medium heat, skimming the surface carefully as the froth rises. Reduce heat to a simmer. Continue skimming until froth no longer forms and you've removed all of it, 10 to 15 minutes. Turn heat to low, add salt and bay leaves, cover pot and let the chicken cook slowly until very tender, 45 to 60 minutes.

Meanwhile, prep the vegetables. When chicken is done, lightly rinse the veggies and add them to the pot. Return to a simmer and cook until vegetables are tender, 12 to 18 minutes. If it looks like the water is getting too low for "soup," add more.

Add the dill, parsley, lovage or celery leaves, and noodles to the soup. Cook until the noodles are barely tender and they "swim to the top," another 5 to 7 minutes.

Stir in lemon juice and add salt to taste. Turn off heat and let soup sit for 30 minutes or so before serving. Reheat it just before serving. Serve each bowlful with a piece of chicken still on the bone, and garnish with a dollop of sour cream and a sprinkling of dried chile flakes, as desired.

David Flesch

HOOK'S CHEESE AND VEGETABLE SOUP
(AND VARIATIONS)

4 TO 5 SERVINGS

The list of reasons why Hook's Cheese is truly Something Special from Wisconsin™ is lengthy. Like other members of the state's program to promote real-deal Dairyland products, Hook's makes top-notch specialties that feature local ingredients—in their case, more than 60 cheese varieties whose milk comes from small farms within a few miles of the factory.

Best known for full-flavored cheddars and creamy blue cheeses, the small operation, run by husband-and-wife team Julie and Tony Hook, is also distinguished for its unusual location among the historic buildings of downtown Mineral Point. Julie is one of the few female cheesemakers in the state, and Tony has been making cheese since he graduated from high school in 1970. And then there are the awards—lots and *lots* of them over the years, including more than a dozen for their sweet-tempered Little Boy Blue, and the one for Colby that made Julie the first woman to win the World Championship Cheese Contest.

Perhaps most amazing is that these shining stars of the cheese-sphere have not let success go to their heads. Their prices remain reasonable. Despite a robust wholesale market, they still come to the farmers market week in and week out to sell directly to customers. Their stand is typically jammed with passersby who, even if they didn't know the caliber of Hook's products, would stop to taste the smorgasbord of cheese samples the couple generously shares.

Like the Hooks themselves, their veggie-and-cheese soup is unassuming and easy to like. It's also easy to vary, so we came up with a couple of adaptations that give multicultural appeal to the basic recipe (see pages 50 and 52), and we encourage you to do likewise. If you're not lucky enough to have the full array of Hook's Cheese available to you, feel free to substitute with similar varieties from other excellent Wisconsin cheese companies.

1 cup chicken or vegetable stock

2 cups chopped broccoli

1 1/2 cups thinly sliced carrots

3 tablespoons butter

1/2 cup chopped onion

3 tablespoons flour

1/2 teaspoon ground nutmeg

1/4 teaspoon black pepper, plus more to taste

3 cups whole milk

2 cups shredded Hook's Colby cheese

1 cup shredded Hook's Swiss cheese

Salt, if needed

Bring the stock to boil in a soup pot over medium-high heat. Add broccoli and carrots, bring to a simmer, reduce heat to low, cover and cook until veggies are nearly tender, 5 to 7 minutes. Drain veggies through a strainer set over a bowl; reserve liquid.

Add the butter to the pot and melt it over medium heat. Add onions and cook, stirring occasionally, until tender, 4 to 5 minutes. Use a wooden spoon to stir in flour, nutmeg and pepper. Stir and cook mixture 2 to 3 minutes.

Warm the milk in a microwave. Whisk the warm milk and reserved liquid into the onions. Bring to a simmer and cook, stirring often, until soup thickens. Cook 1 to 2 minutes longer.

Turn heat to low and add the cheeses a handful at a time, stirring well after each addition. Stir in the cooked vegetables and let them heat through. Taste and add more pepper and some salt as you like. You can serve it immediately or turn off the heat and let the soup gain more flavor as it cools. Reheat to serve, taking care not to let the soup boil, lest it curdle.

HOOK'S CHEESE SOUP WITH TEX-MEX SAVOR

4 TO 5 SERVINGS

If you've got leftover pulled pork (recipe on page 120) or Mexican-style chorizo from last night's taco party, add it to this soup for extra points with the meat lovers at your table.

1 cup chicken or vegetable stock

1 to 1 1/2 cups chopped poblano peppers

2 cup fresh or frozen and thawed corn kernels

3 tablespoons butter

1 cup chopped red onion

3 tablespoons flour

1 teaspoon cumin (freshly ground, if possible)

1 teaspoon dried oregano

1/4 teaspoon black pepper, plus more to taste

3 cups whole milk

2 cups shredded Hook's Seven Year Cheddar cheese (or other Hook's aged cheddars)

1 cup shredded Hook's Pepper Jack, Goat Pepper Jack or Sheep Pepper cheese

Salt, if needed

Bring stock to boil in a soup pot over medium-high heat. Add poblanos and corn, bring to a simmer, reduce heat to low, cover and cook until veggies are nearly tender, 3 to 4 minutes. Drain veggies through a strainer set over a bowl; reserve liquid.

Add the butter to the pot and melt it over medium heat. Add onions and cook, stirring occasionally, until tender, 4 to 5 minutes. Use a wooden spoon to stir in flour, cumin, oregano and pepper. Stir and cook the mixture 2 to 3 minutes.

Warm the milk in a microwave. Whisk the warm milk and reserved liquid into the onions. Bring to a simmer and cook, stirring often, until soup thickens. Cook 1 to 2 minutes longer.

Turn heat to low and add the cheeses a handful at a time, stirring well after each addition. Stir in the cooked vegetables and let them heat through. Taste and add more pepper and some salt as you like. You can serve it immediately or turn off the heat and let the soup gain more flavor as it cools. Reheat to serve, taking care not to let the soup boil, lest it curdle.

HOOK'S CHEESE SOUP WITH SCANDINAVIAN SAVOR

4 TO 5 SERVINGS

The recipe tester for this soup remarked that "it's like a grilled Swiss cheese sandwich on caraway rye in a bowl." We say: Take that caraway rye, spread it with softened butter, top it with smoked whitefish, thin-sliced red onion and chopped dill, and serve it with this soup.

1 cup chicken or vegetable stock

2 cups finely chopped parsnips

1 cup finely chopped carrots

1/2 cup finely chopped celeriac

3 tablespoons butter

1 cup chopped leeks (white and pale green parts only)

3 tablespoons flour

1 teaspoon caraway seeds, freshly ground with a mortar and pestle or very finely chopped

1/4 teaspoon black pepper, plus more to taste

3 cups whole milk

2 cups shredded Hook's Two-Year Swiss

1 cup shredded Hook's Sheep Milk Dill Havarti

1 to 2 tablespoons chopped fresh dill

Salt, if needed

Bring the stock to boil in a soup pot over medium-high heat. Add parsnips, carrots and celeriac, bring to simmer, reduce heat to low, cover and cook until veggies are nearly tender, 12 to 15 minutes. Drain veggies through a strainer set over a bowl; reserve liquid.

Add the butter to the pot and melt it over medium heat. Add leeks and cook, stirring occasionally, until tender, 5 to 7 minutes. Use a wooden spoon to stir in flour, caraway and pepper. Stir and cook the mixture 2 to 3 minutes.

Warm the milk in a microwave. Whisk the warm milk and reserved liquid into the leeks. Bring to a simmer and cook, stirring often, until soup thickens. Cook 1 to 2 minutes longer.

Turn heat to low and add the cheeses a handful at a time, stirring well after each addition. Stir in the cooked vegetables and let them heat through. Taste and add more pepper and some salt as you like. You can serve it immediately or turn off the heat and let the soup gain more flavor as it cools. Reheat to serve, taking care not to let the soup boil, lest it curdle.

THIS BUD'S FOR YOU

There's only one thing better than getting fresh, first-rate ingredients from the farmers market, and that, of course, is growing them yourself. The benefits that make market shopping so gratifying are also in play when you have a garden: knowing where your food comes from and how it was grown; choosing the types and variety of crops you like best; enjoying them in season, at the peak of ripeness. When you grow and cook your own food, you savor it in a deeper way, because of the effort you've put in. You make an intimate connection with nature, with what really makes life tick. You feel proud and uplifted, like Tom Hanks' character in "Cast Away," when he creates a blaze with two sticks of wood, and joyously roars, "I made fire!"

If you're looking to get your own garden going, the farmers market can help. In late winter, when vendors are sprouting seeds in greenhouses for the coming season, many of them plant extra rows in order to sell bedding plants at market. Indeed, the early season at market is known for its wide variety of annuals, perennials, herbs, vegetable, fruit and flower transplants. Who can resist? Those tables- and shelves-ful of budding greenery are a promise of joy to come.

The beauty here is that growers select varieties that fare well in their own region and they know the varieties that their customers prefer. They're also happy to answer your questions about best growing practices. Buying locally grown bedding plants means you're less likely to spread disease from plants that are shipped from afar. And it's an excellent way to support local farmers during the weeks when their own crops aren't yet ready for sale.

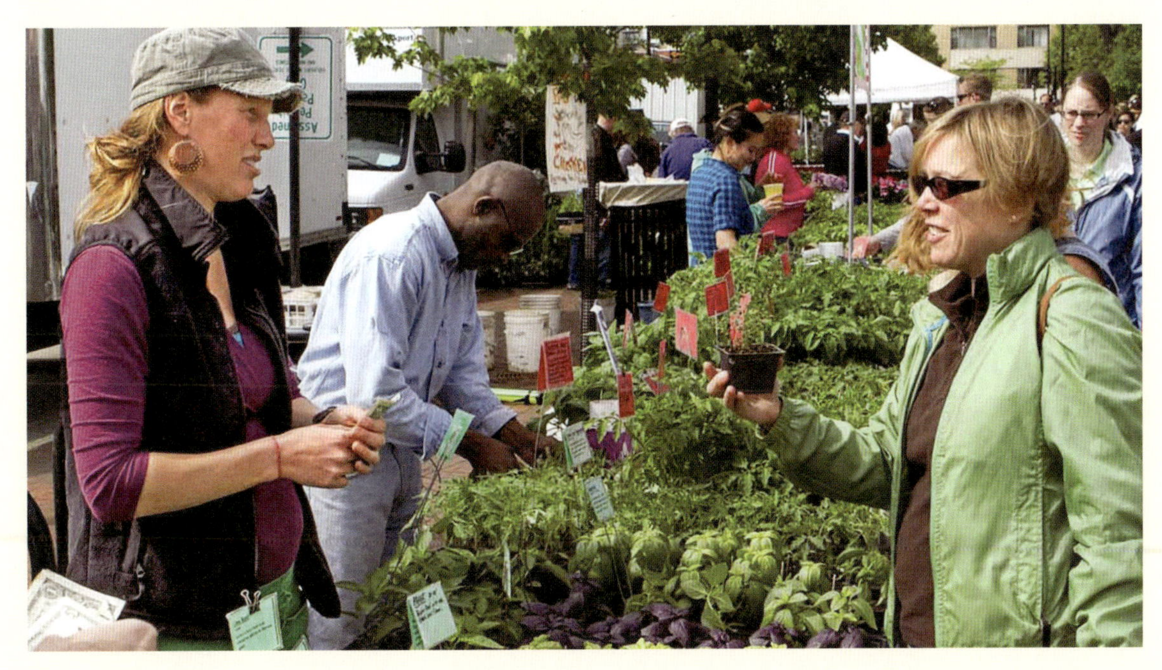

NORTH AFRICAN VEGETABLE SOUP WITH HARISSA YOGURT

6 MEAL-SIZED BOWLS OR 8 APPETIZER SERVINGS

Joan and Ted Ballweg, owner-operators of Savory Accents, market a wide assortment of fresh chiles, ranging from mild shishitos to the hotter-than-you-know-what Carolina Reaper—as well as bedding plants, dried seasonings, chile sauces, dressings, dips, baking mixes, and even chile-spiked sweets like honey, maple syrup and jam.

One of their feistiest products is harissa paste, a seasoned chile condiment that's widely used in the North African countries of Tunisia, Algeria and Morocco. While living in France years ago, Joan fell in love with harissa, and now makes a spunky yogurt sauce with it to top this vegetable-happy soup. Harissa-spiked yogurt is also righteous as a dip for raw vegetables, pita triangles or crackers. (Make a double batch; you'll be glad for the leftovers.)

A little harissa paste goes a long way—you'll only need a teaspoon of it for the sauce. As for the rest of the jar, the paste does wonderful things to eggs, hummus, chili, grilled meats and a whole lot more.

Jackie Hass

Yogurt sauce:

1 cup Greek yogurt

1 tablespoon olive oil

1 teaspoon bottled harissa paste

1 teaspoon minced garlic

Coarse kosher salt to taste

Vegetable soup:

1 pound carrots, peeled and cut into 1/2 to 3/4 inch pieces (about 3 cups)

1 turnip or rutabaga, peeled and cut into 1/2 to 3/4 inch pieces (about 3 cups)

3 tablespoons butter

1/2 pound spring onions or green onions, coarsely chopped (about 1 3/4 cups)

2 tablespoons chopped flat-leaf parsley

2 tablespoons chopped fresh mint

2 to 3 teaspoons minced garlic

1 teaspoon paprika

1 teaspoon ground cumin

1/2 teaspoon ground coriander

Salt and pepper

1/2 cup dry white wine

2 teaspoon flour

1 can (15 ounces) garbanzo beans

1 bag (5 ounces) baby spinach (about 6 loosely packed cups)

1 to 3 teaspoons fresh lemon juice

Whisk all yogurt sauce ingredients in a medium bowl. Cover and chill. (This can be done up to two days ahead.) Prepare a large bowl of ice water.

For the soup, bring 8 cups water to boil in a large saucepan over high heat. Add carrots and cook until just tender, 4 to 6 minutes. Use a large slotted spoon to transfer carrots to ice water. Return cooking liquid to boil. Add turnip or rutabaga and cook until just tender, 3 to 5 minutes. Transfer them to the bowl of ice water. Reserve cooking liquid.

Melt the butter in a large, heavy pot over medium heat. Add onions, parsley, mint, garlic, paprika, cumin and coriander. Sprinkle lightly with salt and pepper. Cook until onions are soft, stirring often, 5 to 8 minutes. Add wine and simmer until liquid is reduced by about half, approximately 5 minutes. Stir in flour.

Add cooked vegetables, beans, spinach and 4 cups of the reserved cooking liquid. Bring to a simmer and cook until vegetables are heated through, adding more cooking liquid as desired.

Season soup to taste with salt, pepper and lemon juice. You can serve it right away or let it develop more flavor by cooling it down and/or refrigerating it for several hours to overnight, then reheating just before serving. Top each bowlful with a dollop of harissa yogurt sauce.

BULGARIAN COLD CUCUMBER SOUP

4 TO 6 SERVINGS

Contributor Betsy Abramson has been a regular customer of the DCFM for most of her 50 years living in Madison. She says, "I enjoyed going there with just my husband, enjoyed going with my kids, enjoyed taking out-of-town guests—and now I enjoy taking my grands."

Betsy got this from her mother, who got it from her friend Helen ... and as is often the case with handed-down recipes, it's been modified along the way (including by us). It's a fabulous cold soup with a great mix of textures—the smoothness of the yogurt and the crunch of the diced cucumbers and walnuts. It's also incredibly refreshing on a humid summer day. "Fresh dill is ESSENTIAL," says Betsy. She also gives it flair by double-chilling it—once in the refrigerator and then again just before serving, when she stirs in ice cubes that both thin the soup and make it extra cold. (Talk about perfect for a picnic!)

1 1/2 cups peeled, seeded and finely diced cucumbers

1/2 to 1 cup chopped walnuts, lightly roasted

2 tablespoons chopped fresh dill

Scant 1 teaspoon salt

1/4 teaspoon black pepper

2 tablespoons extra virgin olive oil

1 good-sized clove garlic, minced, lightly salted and pressed to a paste with the flat of a knife

1 1/2 cups plain whole milk yogurt

4 to 6 ice cubes

Terese Allen

Mix everything except ice cubes in a mixing bowl. Cover, place in fridge and chill thoroughly until ready to serve (up to overnight).

Before serving, transfer mixture to a large serving bowl (glass, if you have one). To serve: Bring the bowl to the table, add ice cubes to soup, and stir, stir, stir, stir. It should have the consistency of chilled borscht. Ladle soup into serving bowls, including what's left of one ice cube per bowl. Encourage guests to keep stirring until ice is completely melted. Then enjoy!

SALADS

INDIGENOUS SALAD

4 TO 6 SERVINGS

Oneida Nation member and Yowela Farms grower Dan Cornelius uses regenerative methods that emphasize soil improvement to cultivate native varieties of corn, beans, squash and other vegetables. He raises grass-fed cattle, poultry and goats; he hunts game, taps maple trees and harvests wild rice. Dan also teaches college courses that focus on Wisconsin's historic and contemporary Indigenous foodways. When there's spare time in his demanding schedule, Dan sets up a stand at the DCFM, offering traditional ingredients like dried hominy and authentic, hand-harvested wild rice.

"Cornelius is trying to farm the way his ancestors did in hopes that future generations can feed themselves," the Wisconsin State Journal's Chris Hubbach has written. "He's part of a small but growing 'food sovereignty' movement of people seeking to provide themselves with healthy, traditional foods produced through sustainable practices that can help solve problems like water pollution and climate change."

Dan shared this recipe for a delectable salad that melds indigenous ingredients, ancient values and modern flair. You can use most any variety of cooked fresh shell beans here; cooked dried beans may also be substituted. Note that the wild rice, hominy and shell beans called for must be pre-cooked; see sidebar on page 59 if you need guidance for that. And don't be afraid to make extra; once cooked, all three of these native foodstuffs freeze well.

Dressing:

2 tablespoons apple cider vinegar

2 teaspoons maple syrup

1 scant teaspoon minced garlic

1/2 teaspoon salt, plus more to taste

1/4 teaspoon pepper, plus more to taste

2 tablespoons sunflower or olive oil

2 tablespoons minced fresh herbs, like parsley or dill (optional)

Salad:

2 cups cooked wild rice

1 cup cooked hominy

1 cup cooked fresh shell beans such as borlotti or cranberry beans (see sidebar on page 59)

1/2 cup fresh blueberries or other seasonal berries

1/2 cup finely chopped apple

For the dressing: Mix vinegar, maple syrup, garlic, salt and pepper in a large bowl. Whisk in the oil in a thin stream. Stir in the herbs, if using. Stir in salad ingredients. Add additional salt and pepper to taste. Serve chilled or at room temperature.

COOKING WILD RICE

Shutterstock

True wild rice is more nutritious, flavorful and delicate than paddy-raised, and it takes less time to cook. Rinse it well, place in a pot and cover it by 2 or 3 inches with cold water. Bring it to a boil and reduce to a simmer. (No need to cover the pot tightly, but you can leave a lid on it ajar to help control the simmer.) Start checking for doneness after about 20 to 25 minutes. You want tender and chewy here, not mushy. When it's ready, strain the rice over a bowl and use the leftover liquid as a hot tea or a stock for soup.

COOKING HOMINY

Shutterstock

Hominy is dried, large-kernel corn that's been nixtamalized—that is, treated with an alkaline solution to loosen the hulls and tenderize the kernels. To cook it, start by soaking the corn in plenty of cold water for 6 to 10 hours. Drain the corn, place it in a pot and cover it by 2 to 3 inches with fresh water. Bring it to a boil over high flame and reduce to a simmer. (No need to cover the pot tightly, but you can leave a lid on it ajar to help control the simmer.) Depending on the type and age of the kernels, it will take 1 to 3 hours of simmering to reach ideal doneness. If necessary, add more water as it cooks to keep the kernels covered. Hominy should be tender all the way through but with some pleasant chewiness. Drain and discard the liquid.

COOKING SHELL BEANS

Shutterstock

Shell beans are harvested for the seeds rather than the pods. Bean seeds can be removed from the pod when they're fresh and then cooked, but they are most often dried in the pod, shucked and then stored for later use. You're probably used to the latter but not so much the former, which is a shame because fresh shell beans are a wonder—they're plump, moist and sweetly vegetal, and they soak up prominent flavors deliciously (alliums, herbs, chiles, smoked or cured meats, olive oil, etc.).

Fresh shell beans cook much faster than dried. Favas and edamame, for example, take only a few minutes of simmering. To cook borlotti, cranberry beans and similar varieties, start by placing in a pot and covering them by an inch with water. Bring the liquid to a boil over high heat, then quickly reduce it to a simmer. Cook—uncovered or partially covered—until the beans are creamy-tender all the way through. This could take as little as 20 minutes or as much as 40, so check them often and don't let them get mushy. For full creaminess, turn off the heat and let them cool in their liquid. The liquid makes an excellent stock for soups, stews and pilafs.

CABBAGE SALAD WITH SOY-GINGER DRESSING

8 TO 10 SERVINGS

Lisa Dussault is an excellent cook whose meals often lean Southeast Asian because her husband once lived in Thailand. She's also one of the early bird regulars at the Capitol Square market, the type whose attention is so laser-focused on the produce that you can almost see the meal-planning wheels churning in her brain. If you want to get Lisa's attention, tap her on the shoulder and ask what she'll do with all the vegetables in her bags. Her eyes ignite and she launches immediately into several possibilities—and they're typically something colorful and vibrantly flavored, like this spicy cabbage salad.

You can use any type of cabbage for this recipe—green, red or purple, Napa, Savoy, etc. She gets all the veggies, as well as the ginger and chili oil, from the DCFM. The salad goes well with grilled chicken or pork (try the Grilled Pork Tenderloin with Cilantro Stems and Black Pepper Rub, page 122). You can also use the dressing as a sauce for stir-fries.

Dressing:

1/2 cup avocado oil or other neutral oil

2 tablespoons rice vinegar (or more if you like things on the tart side)

2 tablespoons tamari or soy sauce

1 heaping tablespoon grated fresh ginger

1 tablespoon sugar

1 teaspoon Kingfisher Farm Szechuan Oil or other bottled Asian-style chili oil

Salad:

8 cups thinly sliced cabbage

1 small bunch green onions (white and light green sections only), finely chopped

1 large carrot, cut into thin strips

1 medium purple daikon radish, cut into thin strips

Leaves from 1 small to medium bunch cilantro, chopped

Garnish (optional): Sesame seeds

Add the dressing ingredients to a bowl and stir or whisk to combine. Let the mixture stand while you prepare the vegetables.

Toss vegetables and cilantro in a large bowl. Add most of the dressing and toss again. Taste the salad and, if you like, add the rest of the dressing and toss once more. Garnish with sesame seeds.

Arthur Wieczorek

Lucinda Ranney

QUICK CURTIDO
(SALVADORAN CABBAGE SLAW)

ABOUT 8 HALF-CUP SERVINGS

Curtido is a Salvadoran cabbage slaw that is often served with *pupusas*, the stuffed corncakes that are a national favorite in El Salvador. "I love to make curtido when the market is overflowing with fresh produce in the middle of the summer," says contributor Kristin Korevec. "Hello, giant cabbages!"

"This version is a quick-pickled slaw, similar to refrigerator pickles. When I crave the acidity of fermented foods but need something quicker and easier, it's my go-to recipe." Kristin serves curtido on tacos, salads, and sandwiches, with eggs, or as a quick snack while she's making dinner. "I often double the recipe since it keeps so well in the refrigerator. And you can really make it your own. Swap in a red onion for the white onion, or purple cabbage for the green; add more or different peppers for additional heat, etc."

Kristin has been attending the DCFM religiously for more than two decades. "I go because of the relationships I've developed with vendors, the incredible array of fresh produce and the unique location on the Capitol Square. It's the only way I know how to start my weekend."

1/2 medium green cabbage, finely shredded (about 4 cups)

1/2 large white onion, thinly sliced (1/2 to 3/4 cup)

2 medium carrots, peeled and grated (about 1 1/2 cups)

1 jalapeño, sliced

1/2 cup apple cider vinegar

2 teaspoons kosher salt

2 teaspoons dried oregano

1/2 teaspoon freshly ground black pepper

"I go because of the relationships I've developed with vendors, the incredible array of fresh produce and the unique location."

Bring 4 cups of water to a boil. Add the cabbage and onion to a large bowl. Pour the hot water over the cabbage and onions until they are submerged. After about two minutes, drain the cabbage and onions, rinse with cold water, then press out as much of the water as possible with your hands. The cabbage and onions should be slightly softened.

Return cabbage and onions to bowl. Add carrots and jalapeño; toss lightly to combine. Add vinegar, salt, oregano and pepper; toss to combine. Store in an airtight glass container or jar in the refrigerator.

You can serve it after about one hour of curing, but it will soften and become more flavorful the longer it cures. Store in the refrigerator for up to a month.

PANZANELLA SALAD

4 TO 6 SERVINGS

This iconic Tuscan salad is at its best when tomatoes are at their cheekiest—fat, fresh and summer-sun-ripened. But it's still wonderful when what's available is hydroponically or hoophouse-raised by local farmers. That means you can have your tomato salad and eat it, too, most any time of the year.

Often served as a first course, the combination of bread and tomatoes is substantial enough for the salad to stand alone as a lunch entrée. Add your favorite grilled chicken breast, steak or shrimp, and call it dinner. Leftover dressing can be used on other salads.

The recipe is slightly adapted from one sent by Dino Maniaci of D'Vino restaurant in Madison. Chef Dino gets his heirloom cherry tomatoes for the salad from longtime vendor Mary Uselman of Don's Produce (named for her late husband, Don Uselman). "Dino's a great customer who has been buying from us for years," says Mary.

1/4 cup balsamic or red wine vinegar

1 tablespoon dried oregano

1 teaspoon salt

1/2 teaspoon pepper

Scant 1/2 cup extra-virgin olive oil, or more if needed

4 cups heirloom cherry tomatoes, halved

1/2 red onion, sliced crosswise paper-thin (about 3/4 cup)

4 cups Homemade Croutons (see sidebar on page 66), or more as desired

4 to 5 cups arugula or other salad greens

Garnish (optional): Flake salt and freshly ground black pepper

Whisk vinegar, oregano, salt and pepper in a medium bowl. Keep whisking constantly while you slowly add a scant 1/2 cup of olive oil. Taste, and if it's vinegary, whisk in more oil until the flavor is balanced—bright-tasting and slightly sharp, but not stingingly so. Add a bit more salt and pepper at this point, too, if the overall taste seems flat. When the dressing is right, add tomatoes and onions to the bowl; stir gently to coat them. Allow mixture to marinate at least 1 hour before serving.

About 2 to 3 minutes before serving, use a slotted spoon to transfer tomatoes to a large salad bowl. Add croutons and some of the dressing; toss well. Add more dressing, if needed, but take care not to soak the croutons. Serve the salad just as the croutons have started to absorb dressing but still have a slight crunch to them. Pile the salad atop arugula or other greens on individual serving plates or on a platter. Offer flake salt and freshly ground pepper to top each salad, if available (or have salt and pepper shakers on the table so folks can season to taste).

You can have your tomato salad and eat it, too, most any time of the year.

Lucinda Ranney

Terese Allen

HOMEMADE CROUTONS

MAKES 4 TO 5 CUPS

Bread needn't be stale to make excellent croutons. Who knew? According to J. Kenji López-Alt, author of *The Food Lab: Better Home Cooking Through Science*, fresh bread that has been oven-toasted has better texture—it's nicely crisp on the outside and tender on the inside, unlike stale bread, which can get tough and leathery when toasted. However, if you have some stale bread on hand that you want to use up, go ahead and use it anyway. The toasting process will refresh the bread enough to work just fine with the flavorful dressing.

2 to 3 tablespoons extra-virgin olive oil

1 teaspoon dried oregano

1 teaspoon dried parsley

1/4 to 1/2 teaspoon garlic salt

1/4 teaspoon dried chile flakes

Good salt and freshly ground black pepper

Baguette-style bread (fresh or stale), cut into 1-inch-thick slices and torn into 1-inch pieces (4 to 5 cups total)

Heat oven to 400 degrees. Whisk olive oil and all seasonings in a medium-sized mixing bowl. Add the bread pieces and toss well. Spread them out in a single layer on an ungreased baking pan. Bake until golden brown, 10 to 12 minutes, tossing them halfway through and watching them carefully to prevent burning. Let cool.

SWEET N' SPICY WATERMELON SALAD OR SALSA

10 TO 15 SERVINGS AS A DIP OR 6 TO 8 SERVINGS AS A SIDE SALAD

To Leslie Ann Busby-Amegashie, this recipe—and others that she has collected into a cookbook about Caribbean cuisine—are a connection to home. She is from Trinidad and moved to Madison in 1994. "The recipes help keep my heritage and culture alive, especially in summer," says Leslie. "I can't go home in summer, but I can bring a taste of home to Madison."

If you want a mild salsa, use one or two jalapeños, but if you like it spicy, go for three or four ... or even more. Be sure to taste the jalapeños first so you have an idea of how spicy they are, and go from there.

This recipe is slightly adapted from Leslie Ann's cookbook, *A Pinch of Dis, A Pinch of Dat.*

3 cups finely chopped seedless watermelon (for salsa) or 3 cups bite-size pieces (for side salads)

2 cups finely chopped cherry tomatoes (for salsa) or 2 cups quartered cherry tomatoes (for side salads)

1/2 cup finely diced red onion

3 finely chopped green onions (about 1/3 to 1/2 cup)

1 to 4 seeded and minced jalapeños

1/4 cup chopped fresh cilantro

2 tablespoons fresh lime juice

1 to 2 teaspoons minced garlic, or more to taste

Salt and pepper to taste

Garnish (optional): Cilantro sprigs

Tortilla chips (optional)

Lucinda Ranney

"The recipes help me keep my culture and heritage alive..."

Combine all ingredients except cilantro sprigs in a medium bowl and mix well. Cover and refrigerate for at least 1 hour, gently stirring it occasionally. When the salsa or salad is chilled, scoop it into a serving dish and garnish with cilantro sprigs. Enjoy the salsa with tortilla chips.

FLOWER POWER

If the farmers market was a cake, flower stands would be the icing on it. Colorful, fragrant and ineffably tempting, a Saturday morning display of gorgeous blossoms make human beings feel instantly happy. Who wouldn't want to tuck a bunch or two inside their shopping cart? Plus, similar to food purchased directly from growers, market blossoms are often fresher, longer lasting and less expensive than what commercial florists can offer. Buying a local bouquet is a way to support area farmers and put less stress on the environment. "Locally grown flowers also carry the story of the people behind them—generations of farmers nurturing something beautiful to be enjoyed and shared by their community," wrote Brie Mazurek of Foodwise, a farmers market organization in San Francisco.

Lucinda Ranney

What's more, in many cases you can have your flowers and eat them, too. Edible blooms add whimsy and bursts of flavor to market-based meals. You might stuff squash blossoms with goat cheese, or toss nasturtiums with salad greens. You can garnish plates with sweet, vegetal scarlet runner bean or pea blossoms. You'll taste a hint of cloves in fuchsia-colored dianthus or discern mintiness in spiky-leaved bee balm. You'll smell citrus when you touch diminutive signet marigolds with names like Lemon Gem and Tangerine. At the market, look for all these and more, like spicy chive flowers, deep blue bachelor buttons, calendula, pansies and johnny-jump-ups.

Edible flowers are delicate, though, so take care. Keep them cool in airtight containers, with a piece of damp paper towel placed beneath the petals, and most types will last several days. And remember: Only some flowers—and flower parts—are safe to eat. Here's where having the grower right across the display table comes in handy. To learn more, just ask the flower vendor.

VIETNAMESE RICE SALAD WITH GREEN VEGGIES AND FRESH HERBS

4 MAIN COURSE SERVINGS OR 6 TO 8 AS A SIDE SALAD

Recipe creator Megan Bjella writes, "While I am no authority in Vietnamese flavors or cooking techniques, I can say with confidence that this rice dish makes a super tasty meal, especially on a hot day. The marinade is essentially a recipe for *núóc châm*, a funky, salty-sour-sweet Vietnamese sauce. Marinating the veggies in it mellows their 'rawness' while maintaining the refreshing flavor and texture of a fresh salad. It makes a satisfying supper when topped with some sort of protein. We've had it with eggs, whitefish and canned sardines. Shrimp or marinated beef or tofu would also be tasty, I bet."

Marinade:

1/4 cup lime juice

2 tablespoons grated fresh ginger

Sliced jalapeño to taste (seeded or not, depending on your heat level preference)

1 tablespoon bottled fish sauce

1 clove roasted garlic, mashed to a paste (minced fresh garlic may be substituted)

1 teaspoon sugar

1 teaspoon black pepper

Salad:

2 cups thinly sliced celery or 2 to 3 cups seeded and sliced cucumbers (peeled or unpeeled)

1 to 2 sliced green onions

1 1/2 cups jasmine rice

1 cup peas, fresh or frozen and thawed

2 loosely packed cups chopped mint, basil and/or cilantro

1/2 cup coconut milk

Salt

Garnish (optional): Roasted peanuts or cashews, chopped

Combine all marinade ingredients in a large bowl. Add the celery or cucumbers and green onions; toss gently and let stand 1 or more hours, or up to overnight, tossing occasionally. (Refrigerate them if it will be a few hours or longer.)

Combine rice and 2 1/4 cups water in a saucepan; bring to a boil over medium-high heat. Reduce heat to low, cover pan tightly and cook rice 15 minutes without lifting the lid. Remove from heat, uncover the pan, place a dish towel over the rice, replace the lid and let stand 10 minutes. Uncover rice, fluff it with a fork and let cool for 15 minutes or so.

Combine marinated vegetables (including their liquid), peas, chopped herbs and cooled rice in a large bowl. Add coconut milk and toss. Taste and season as needed with salt. Serve with chopped peanuts or cashews sprinkled on top, if desired.

"While I am no authority in Vietnamese flavors or cooking techniques, I can say with confidence that this rice dish makes a super tasty meal..."

WINTER SLAW WITH DRIED CRANBERRIES AND SHAVED PARMESAN

ABOUT 8 SERVINGS

Cutting raw cruciferous vegetables into thin shreds tenderizes them while retaining their fresh edge. You've likely done this to kale and cabbage, but what about Brussels sprouts? Don't be shy about adding these tiny cabbages—developed in Belgium as far back as the 13th century and perfectly suited to the Wisconsin growing season—to your slaw repertoire. (You know that they taste best when harvested after a light frost, right?)

With its deep green, burgundy and cream hues, this pretty salad would look splendid on a Thanksgiving table. The recipe is an Ina Garten creation, adapted for us by market devotees David Mandehr and Jane McMillan.

10 to 12 medium sized Brussels sprouts (about 6 ounces), trimmed, halved and cored

Leaves from 6 to 8 large kale stalks

1/2 small head radicchio or red cabbage (about 6 ounces)

1/4 cup lemon juice

1 teaspoon kosher salt

1/4 to 1/2 teaspoon freshly ground black pepper

1/2 cup extra-virgin olive oil

1 cup dried cranberries

5 to 6 ounces good-quality Parmesan or other Italian-style hard cheese

Use a sharp knife to cut the Brussels sprouts, kale and radicchio or red cabbage crosswise into thin shreds. Transfer them all to a large bowl.

Make dressing: Place lemon juice, salt and pepper in a small bowl and whisk well to dissolve the salt. While constantly whisking, slowly add the olive oil in a thin stream.

Pour about 1/3 cup of the dressing over the vegetables, just enough to moisten them well, and toss. Add the dried cranberries and toss again.

Use a wide vegetable peeler to shave the Parmesan into big shards. Add them to the salad and toss gently to keep the Parmesan shards intact. Let stand at room temperature for 2 to 3 hours, tossing once or twice during that time.

Adjust the flavorings just before serving, adding more dressing, salt and pepper as needed for a lively, balanced flavor. (Leftover dressing can be used for other salads.) Serve at room temperature or chill the salad for an hour or two first.

SWEET CORN AND BELL PEPPER SALSA AND SALAD

3 TO 4 SERVINGS

Katie Webster developed this recipe when she was an undergraduate at the University of Wisconsin-Madison. "I hadn't been to many farmers markets before coming to Madison and quickly found the corn and bell peppers that were available to be nothing like I had ever had before," writes Katie.

"I made this recipe one day after I 'accidently' bought too much corn and peppers and needed to use them before it was too late. I roasted them and made a salsa that became a favorite of mine, something I would keep in the refrigerator to add to a quick salad or to things like taco boats or burrito bowls. It would also go with nachos and black bean burgers or anything your heart desires.

"Although I tried to make this salsa for my family when I went back home, something just wasn't the same without the fresh Wisconsin produce." (We know what you mean, Katie.)

Katie shares one of the ways she likes to serve her salsa, which, by the way, has no hot chiles in it and, therefore, is perfect for people who *don't* like it hot. Her "salsa salad" features crisp greens, crunchy croutons and a simple dressing.

Lucinda Ranney

Terese Allen

Salsa:

1 sweet bell pepper (red, orange or yellow)

1 large ear sweet corn

1/2 to 1 tablespoon olive oil

1 tablespoon lime juice

2 tablespoons chopped cilantro

Salt and freshly cracked black pepper

Salad:

1 medium head romaine or other favorite lettuce, chopped

Homemade Croutons (see page 66)

1 tablespoon white wine vinegar or balsamic vinegar

Salt

2 tablespoons olive oil

Honey (optional)

Freshly cracked black pepper

Heat oven to 425 degrees. Halve the bell pepper, remove seeds and membranes, and cut each half in two. Rub olive oil onto the pepper pieces and corn cob and place them on a rimmed baking sheet. Roast for 8 minutes, then turn the veggies over. Continue roasting until peppers are just tender, another 3 to 5 minutes. Remove peppers, turn the corn again, and continue roasting the corn until it is partially browned, another 3 to 7 minutes.

As soon as the pepper pieces come out of the oven, place them in a bowl, cover it tightly and let stand 10 to 20 minutes to release the skin from the peppers. Pull off and discard the pepper skins, chop the peppers and place them back in the bowl. When the corn is done and cool enough to handle, cut the kernels off the cob and add them to the peppers.

Stir in the lime juice, cilantro, salt and pepper to taste, and mix until fully incorporated. Store salsa in the refrigerator until ready to use.

To prepare the salad: Wash, chop and dry the lettuce in a salad spinner or clean towel. Transfer it to a large bowl. Add the salsa and whatever number of croutons you like. Whisk vinegar and a little salt in a small bowl. Whisk in the olive oil a little at a time. Toss the dressing with the salad. Drizzle honey over top of the salad, if you like, and top with cracked black pepper.

GRILLED VEGETABLE COUSCOUS SALAD

ABOUT 12 SIDE-DISH SERVINGS (6 TO 8 IF SERVED AS A MAIN COURSE)

Here's a summer salad that melds the exuberant Mediterranean flavors of cumin, garlic and cilantro with grilled market vegetables, chewy beads of Moroccan couscous and chile-infused oil. It'll bring on happy groans as a side dish at potlucks, or as a main course—topped with grilled chicken or tofu skewers—for family dinner. (Make it as much as a day ahead and it even improves in flavor).

The recipe—and the chili oil—comes from Joan Ballweg of Savory Accents. Joan and her co-owner husband, Ted, could rightfully be called the DCFM's hottest couple—they grow a wide variety of chiles, selling them fresh and dried, as well as in hot sauces, infused oils and more.

Cooking note: Couscous is really a kind of pasta, but in miniature form. There are several types available—the one used here is Moroccan couscous. It's the smallest and cooks very quickly. For the chili oil, a great local choice is Savory Accents Chili Oil or Pepperolio.

1 1/4 cups Moroccan couscous

4 small zucchini (about 1 1/4 pounds), cut in half lengthwise

2 medium red, yellow or orange sweet bell peppers (about 3/4 pound), halved and seeded

1 medium globe eggplant (3/4 to 1 pound), cut crosswise into 1/3-inch-thick rounds

1 medium jalapeño or other hot chile pepper variety (optional)

About 6 tablespoons olive oil, divided

2 cups small tomatoes (about 3/4 pound), cut in half

1/3 cup chopped cilantro

1 tablespoon good chili oil, or more to taste

6 tablespoons lemon juice

1 teaspoon ground cumin, or more to taste

1 tablespoon minced garlic, or more to taste

Salt and pepper

Couscous is really a kind of pasta, but in miniature form.

Prepare and heat an outdoor grill using wood, charcoal or gas. (You could also use an indoor griddle.)

Bring 2 1/2 cups water to boil in a medium saucepan. Add the couscous, cover tightly, turn off the heat and let stand 5 minutes. Uncover, fluff couscous with a fork, and let it cool while you grill the vegetables.

Brush zucchini, bell peppers, eggplant and jalapeño (if using) with 2 tablespoons of the olive oil. Grill them over medium-high heat, turning occasionally, until there is some char on their surfaces and vegetables are barely tender. Timing will be different for each type, ranging from about 3 to 4 minutes for the jalapeño, to 8 to 10 minutes for the eggplant. Take care not to let them get overcooked.

Let vegetables cool, then cut jalapeños in half and scrape out the seeds and membranes. Chop the jalapeño finely and add it to a large bowl with the couscous. Cut bell peppers, zucchini and eggplant into bite-sized pieces and add it to the bowl of couscous. Add the tomatoes and cilantro, too.

Whisk the remaining 4 tablespoons olive oil, chili oil, lemon juice, cumin, garlic and salt and pepper to taste in a small bowl. Add it to the vegetable-couscous mixture and toss gently but thoroughly. Taste the salad and add salt and pepper to taste. Add more cilantro, chili oil or other flavorings as desired. Serve right away, or let the salad stand at room temperature for up to an hour to let the ingredients get acquainted.

CALLING ALL ANISE HYSSOP LOVERS (CURRENT AND FUTURE)

When we asked Odessa Piper if she would contribute to this cookbook, she wrote back with an eager "yes," saying that she felt like she had been handed her moment to plug one of the most deserving and delightful ingredients she knows. Odessa, in case you didn't know, is the James Beard Award-winning founder and former owner of L'Etoile restaurant—and Madison's fairy godmother of farm-to-table dining. She's also a very big fan of anise hyssop.

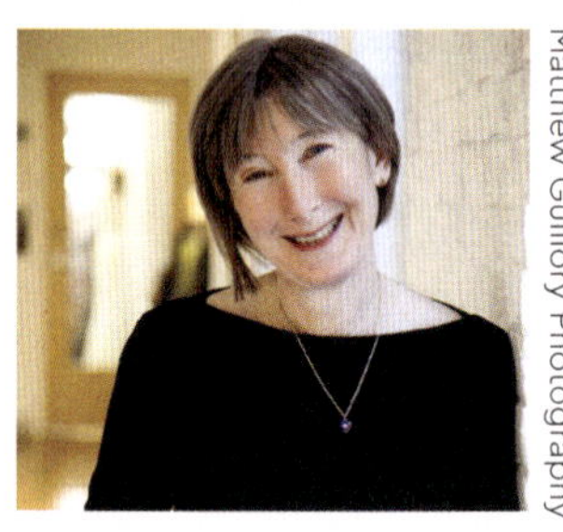
Matthew Guillory Photography

"Anise hyssop is a surprisingly delicious alternative in recipes that call for mint or basil," says Odessa. "The leaves and flowers are mildly sweet and delicately anise-flavored, with a stevia-like effect—meaning it can make tart yogurt or fruits taste sweet without adding sugar. And it works where mint is often too dominant. As for savory uses, I have yet to find a leafy green salad, noodle bowl, vegetable stir-fry, rich chop or roast that it doesn't complement."

At one time, this bodacious flavorant had a brief heyday at the Capitol Square market. Odessa credits grower Rink DaVee for turning her on to it, back when he was fresh from his stint as assistant forager at Berkeley's renowned Chez Panisse. "Vendors used to bring it in as an herb to transplant—also as cut bunches—but it had to compete with all kinds of 'flavored' basils and mints. So its availability at the market has seen better days," Odessa adds. "But anise hyssop is now adored by restaurant chefs, while those fancy flavored basils and mints are a dim memory. I think it could come roaring back."

Indeed it could come back, for in true *Field of Dreams* fashion ("If you build it, they will come"), that's the beauty of the farmers market. If you're a shopper with a special crop to recommend, ask for it, and growers will give it a go. If you're a vendor, put it on display, pass out a few samples, and patrons will fall hard.

And if you're Odessa? Create a list of enticing ways for cooks to spotlight this northern prairie native in internationally focused meals throughout the year (see following page), and anise hyssop may once again gain the attention it deserves.

ANISE HYSSOP PLAYBOOK

- Chop leaves into Mediterranean tabbouleh and other grain salads.
- Combine whole young leaves with fresh basil, dill umbels and chives, in California fashion, to make a resinous herb salad paired with balsamic vinegar and salty, grated hard cheese.
- Wrap individual leaves around slices of melon or nectarine.
- Combine leaves and buds with butter to season carrots, beets, pea pods or parsnip (the French call this *beurre composé*).
- Go Greek: Scatter chopped fresh leaves and flowers over sliced watermelon and feta cheese.
- Layer leaves over a slice of cured ham, roll it up and slice into pinwheels. (Repeat, this time with a slice of ripe pear in the center of the ham slice.)
- Macerate julienned leaves with stone fruits like apricots, cherries and peaches.
- Wrap leaves around batons of good aged cow's or sheep's milk cheese.
- Salute Scotland: Stir chopped leaves into shortbread cookie dough and press the cookie tops with buds and sugar before baking.
- For an updated American favorite, stir chopped leaves into oat-and-butter streusel topping and bake atop apples or rhubarb.
- To make gifts of indigenous tea, dry mature leaves and flowering stems in a basket and then package them into clear bags. Steep about 4 tablespoons in a large tea pot for 8 minutes.
- Macerate leaves in a simple syrup, pulverize and churn into sorbet, serve with scoops of fruit sorbet and ice cream.
- Chew on a leaf after drinking coffee to freshen the breath.

© Shutterstock

GARLIC SCAPE PESTO PASTA SALAD

4 TO 8 SERVINGS

There is no end to the possibilities when you mix pesto, pasta and seasonal vegetables. This combination, from Jackie Gennett of Bushel & Peck's, is a lively salad made from late spring/early summer produce: garlic scapes, snap peas, radishes and green onions.

Cooking note: To get the three cups cooked rotini called for in this recipe, you'll need to boil about three scant cups of dried pasta. Be sure not to overcook pasta when using it in salads.

4 cups cooked rotini or other small, shaped pasta

1/4 cup snap peas, cut into 1/2-inch pieces

1/4 cup sliced French breakfast radishes

1/4 cup sliced green onions

1 1/2 cups Garlic Scape Feta Pesto (see page 182)

Salt and pepper to taste

Garnish (optional): Fresh basil leaves

Toss all ingredients in a large bowl. The flavors will mellow and blend if you make it ahead of time—even the day before—but it can be served right away. To serve, spoon the mixture into a pretty bowl or onto a platter. Garnish with basil leaves, if desired.

ASIAN TOASTED CABBAGE SALAD

5 TO 6 SERVINGS

"I like to think of our family's eating habits as adventurous, and I owe much of that to the DCFM," says former vendor Sue Gronholz. She got this recipe from a neighbor "who thought it sounded 'weird' and knew my family would like it. She was right!"

Sue played with the recipe to create a version that uses Napa cabbage. "Twenty-some years ago, Napa cabbage was not usually found in small-town grocery stores, so I really appreciated being able to find items like this at the DCFM. Best of all, it was homegrown, which tastes better than any store-bought item."

This salad is perfect for those "I need a make-ahead dish" occasions, and it handles substitutions well, too. Use green or red cabbage instead of Napa; replace some of the cabbage with shredded carrots or slivered red peppers; try peanuts instead of almonds, or sunflower seeds instead of sesame seeds.

Dressing:

1/3 cup olive oil

3 tablespoons rice vinegar

Reserved seasoning packet from ramen noodles (see below)

2 tablespoons sugar

1 teaspoon grated lemon zest (yellow part of peel only)

1/4 teaspoon each salt and pepper

Salad:

1 package (3 to 4 ounces) soy-sauce-flavored ramen noodles, crushed into pieces (reserve seasoning packet for the dressing)

1/2 cup sliced almonds

2 tablespoons white sesame seeds

6 lightly packed cups shredded or very thinly sliced Napa cabbage

1/2 to 3/4 cup chopped green onions

To make the dressing, place all dressing ingredients in a jar with a tight lid and shake well to blend.

Heat a skillet over medium-high flame. Add the crushed ramen noodles and toast them, stirring often, until they're lightly colored, about 5 minutes. Transfer noodles to a large bowl to cool. Toast the almonds and sesame seeds in similar fashion, about 3 to 4 minutes, taking care not to let them scorch. Transfer them to the bowl with the noodles to cool.

Add cabbage, green onions and dressing to the bowl and mix everything well. Serve immediately or refrigerate the salad and serve within a few hours.

Shutterstock

Lucinda Ranney

Lucinda Ranney

Lucinda Ranney

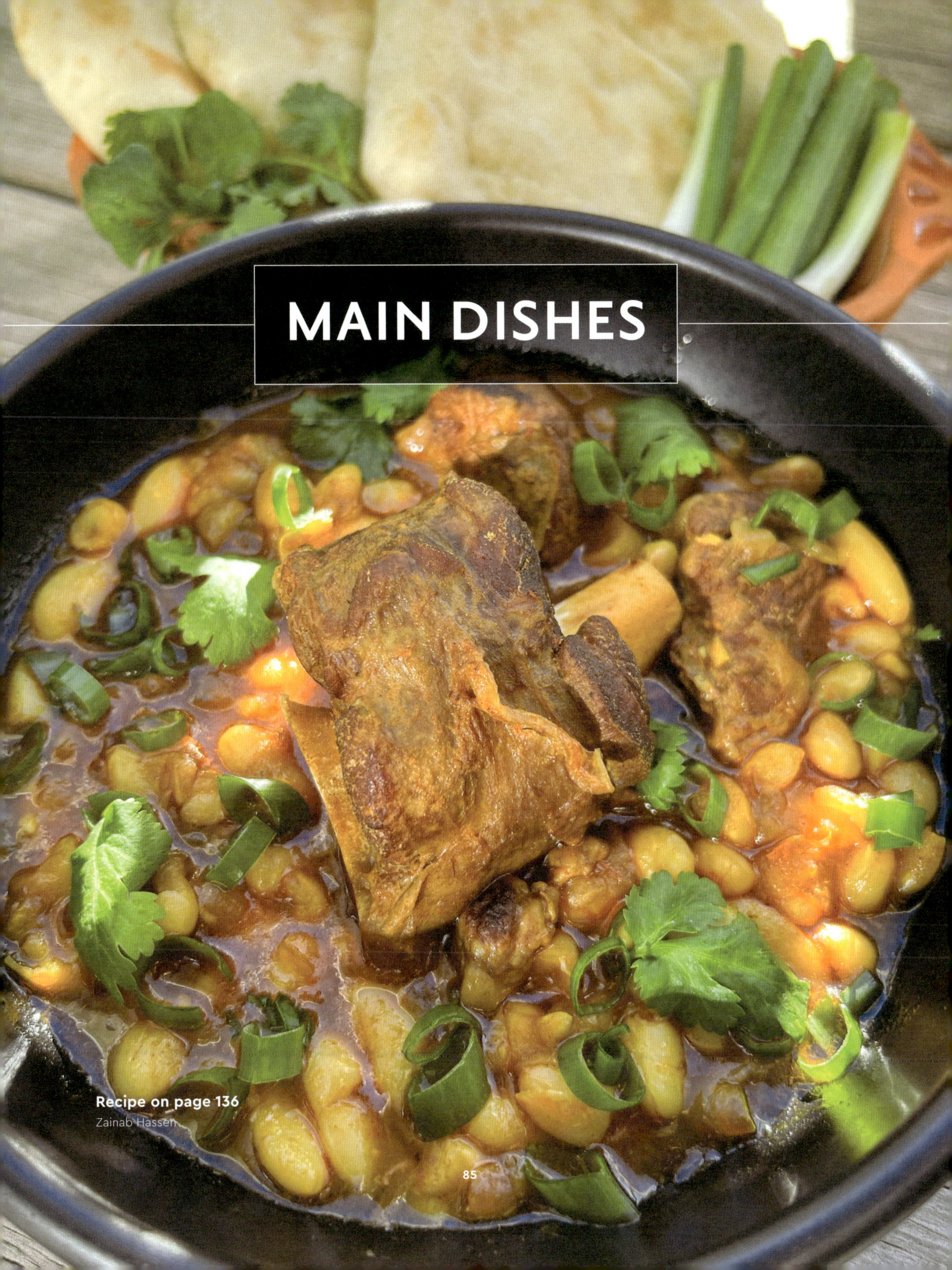

MAIN DISHES

Recipe on page 136
Zainab Hassen

CAULIFLOWER POTATO TARKARI

4 SERVINGS

Tarkari is a spiced—but not necessarily spicy—vegetable curry, says contributor Linda Falkenstein, an editor at *Isthmus*, Madison's alternative newspaper. "I used to lunch quite frequently at Himal Chuli [a well-known Nepali restaurant in town] and loved the daily tarkari. I decided I should be trying to make this at home, so of course I looked to the internet. I ended up smooshing together a couple of recipes, taking the lazy way out at every turn but aiming for fidelity to Himal Chuli's flavor. I find that this cauliflower-potato version comes pretty darn close."

The dish is not only delicious and fairly easy to make, it's also very forgiving and adaptable. "You can also use the base spices and the (possibly inauthentic) simmering method to do the same treatment with any number of veggies besides cauliflower, potato and carrot. Spinach and black-eyed peas are a good [combo] that Himal Chuli features frequently, but you could experiment further. The secret is to not skip the squeeze of fresh lemon at the finish. The dish should be served on brown rice."

Cooking note: Begin cooking the brown rice about 10 to 15 minutes before you start preparing the tarkari.

Ruth Bronston

2 tablespoons butter

1/2 to 1 medium onion, finely diced

1 to 2 teaspoons grated fresh ginger (or 1/2 to 1 teaspoon powdered ginger)

1/4 to 1/2 teaspoon whole coriander seeds

1/2 teaspoon ground cumin

1/2 teaspoon turmeric

1/2 teaspoon salt

1/4 teaspoon black pepper

2 cups finely cubed potatoes such as Yukon Golds (peeled or unpeeled)

2 cups vegetable or chicken stock (or a stock-water combo)

1 cup diced carrots

5 cups cauliflower, cut into 1-inch florets

1 to 2 tablespoons lemon juice

Hot, cooked brown rice

Melt the butter in a large skillet over medium-low flame. Add the onions and cook, stirring occasionally until they begin to soften, 4 to 5 minutes. Stir in the ginger, coriander seeds, cumin, turmeric, salt and pepper and keep cooking until onions are tender, another 2 to 3 minutes.

Raise the heat to medium and add the potatoes and then the stock (or stock-water combo). Bring to a simmer and cook until potatoes are tender, about 10 minutes.

Stir in the carrots and cauliflower, return the mixture to a simmer, then reduce the heat to medium-low and cover the pot. Keep cooking the mixture, removing the lid every so often to stir and check for doneness. Add a little water, if necessary, for a "saucy" dish. It may take up to another 15 minutes of cooking, depending on whether you prefer the cauliflower crisp-tender or very tender.

Finish the dish by stirring in lemon juice to taste. Serve with brown rice.

"I ended up smooshing together a couple of recipes, taking the lazy way out at every turn but aiming for fidelity to Himal Chuli's flavor."

CARAMELIZED RATATOUILLE

4 TO 6 SERVINGS

Stew, sauce or spread? Appetizer, salad or entrée? Simmered, stir-fried or baked? Ratatouille can be any of the above, making it one of the most versatile dishes on the planet. Born and beloved in southern France, it's also the perfect Heartland dish in summer, when its signature combo of eggplants, peppers, zucchini and tomatoes is in high season, or in winter, when a frozen batch brings life to pasta, grains, pizza, bruschetta, melts and more.

Contributor Pam Dempsey, who roasts her ratatouille to caramelize the vegetables (read: more flavor), eats it on its own, with rice or yogurt on the side, or sometimes with yogurt *in* the dish itself. What's more, she thinks it's even tastier after it's been frozen. Did somebody just say "double batch"?

8 tablespoons olive oil, divided

3 medium globe eggplants (peeled or unpeeled)

3 medium sweet red peppers, seeded

3 medium zucchini (a mix of green and yellow is nice)

4 medium tomatoes, cored

2 to 3 medium yellow onions

6 cloves garlic (don't use garlic powder or garlic salt)

1/2 cup finely minced basil leaves (parsley or dill can also be nice additions)

Salt and pepper

Lucinda Ranney

Heat oven to 350 degrees. Use 2 tablespoons of the olive oil to oil two large, heavy, rimmed baking sheets. Cut eggplant, peppers, zucchini and tomatoes into 1/2-inch cubes. Finely chop the onions and mince the garlic. Gently toss the vegetables, garlic, herbs and remaining 6 tablespoons olive oil in an extra-large bowl until well coated.

Spread mixture out evenly in a single layer on the two baking sheets. Bake 30 minutes, then remove pan from oven and gently toss mixture. Bake for another 30 minutes. Transfer to a large serving dish and add salt and pepper to taste. Serve hot, warm or at room temperature.

EGGPLANT APLENTY

When you hear the word eggplant, usually the first things that come to mind are eggplant Parmesan and ratatouille. Too often, these dishes are also the last. A member of the nightshade family, eggplant must have some residual notoriety left over from medieval times, when Europeans believed it was poisonous. But its meaty, neutral appeal and velvety mouthfeel make eggplant a rare bird in the vegetable world. Think of it as the poor man's—or vegetarian's—meat, and you've got the starting point for a world of dishes.

The international applications for eggplant are myriad: Hmong Asian stir-fries, Italian pastas, Middle Eastern spreads, Indian curries. Eggplant can star in soups, salads, stews and sandwiches. It can be marinated and grilled, stuffed and baked, boiled, broiled or fried. Multiply the treatments by the burgeoning number of eggplant varieties available from local farmers—in various shapes and colors, with whimsical names like Fairy Tale, Graffiti or Rosa Bianca—and you've got a vast number of options for dinner. For a few unusual creations, try these recipes: Smoky Eggplant Tomato Sauce, page 180; Fasolia (Libyan White Bean and Lamb Stew), page 136; and Red Kabocha Squash Soup with Ginger, Lemongrass and Thai Eggplant, page 40.

And don't forget those touchstone eggplant dishes! Check out the Caramelized Ratatouille, page 88, and Eggplant Parmesan Express, page 92.

CRISPY KALE AND ROASTED SQUASH WITH TAHINI LIME SAUCE

4 SERVINGS

East meets West when a sultry tahini lime sauce accents kale and winter squash, two autumn standards in our neck of the woods. This vegan recipe, inspired by a Bon Appétit recipe and contributed by Megan Bjella of Madison, calls for delicata squash, which has a mild brown-sugar taste and can readily be cut into handsome half-moon shapes. But try it, too, with more familiar butternut squash, or bright-fleshed kabocha, or whatever looks good at the market. (Doesn't all winter squash look good?)

Megan says there's a secret to making creamy tahini sauces and dressings: It's water. "I used to make tahini sauces with olive oil but could never get the texture or flavor quite right—it turned out gloopy and heavy. A small amount of water keeps the tahini from seizing up, creating a creamy, no-fuss sauce."

She also cautions, "You may have to play a bit with the temperature and timing of the kale chips since ovens vary a bit. Err on the conservative side so you don't burn them."

Cooking note: If you use ghee for the fat in this recipe, you can either make it yourself (see page 140) or you can purchase it. Ghee solidifies when it's cool, so you may have to melt it before proceeding with the recipe.

Char Thomson

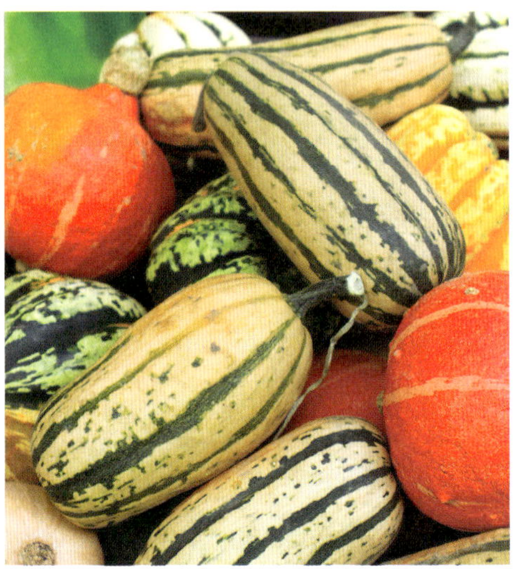

Lucinda Ranney

1 medium bunch curly kale, stems and ribs removed, leaves coarsely chopped (about 5 packed cups)

1 medium Delicata squash, well-scrubbed, cut lengthwise, seeded and cut into 1/2-inch-thick half-moons (about 3 cups)

2 to 3 tablespoons neutral oil (like canola) or melted ghee

Salt

3 tablespoons tahini

2 tablespoons tamari

2 teaspoons honey

2 teaspoons toasted sesame oil

1 tablespoon lime zest

Juice of 1/4 lime, or more to taste

Cayenne or dried chile flakes to taste

Cooked rice or quinoa (optional)

Heat oven to 375 degrees. Line two baking sheets with parchment paper; place kale on one and squash on the other. Drizzle some of the oil or ghee over the kale and sprinkle with salt, then massage kale with your hands until it's bright green and coated with oil. Drizzle more oil/ghee over squash, sprinkle with salt and toss to coat.

Bake kale and squash until kale is crisp, about 12 to 18 minutes, checking halfway through to toss the kale and rotate the baking sheet. Remove kale from the oven when it's done. At this point, turn the squash pieces over and continue roasting until browned on both sides and soft when pierced with a fork, an additional 10 to 15 minutes.

Meanwhile, whisk tahini, tamari, honey, and sesame oil in a bowl. Add lime zest and juice, along with cayenne or dried chile flakes. Slowly whisk in 3 to 4 tablespoons water until sauce is creamy and a little runny but still a little thick. Taste sauce and add additional salt, lime juice or any other of the flavorings as desired.

To serve: Pile kale and squash over quinoa or rice on a platter or individual plates. (Or serve the veggies on their own as a side dish.) Drizzle with about half of the sauce. Serve the remaining sauce in a small bowl for those who want more.

Try it, too with more familiar butternut squash, or bright-fleshed kabocha, or whatever looks good at the market.

EGGPLANT PARMESAN EXPRESS

4 TO 6 SERVINGS

A self-taught cook, Jean Martin has a knowledge of food that has come a long way since she made her first pizza with bread dough, ground beef and American cheese. Today she can advise market-goers as well as a professional chef about cooking with specialty vegetables. Take the eggplant in this recipe: "Over the years I have preferred to use either a round Spanish or Italian eggplant. They are usually light purple or white and purple in color and are less bitter than [standard globe eggplants]. Thus, there's no need to salt and rinse." She thinks that "all that salting started because we weren't getting eggplant from the farmers market! Fresh eggplant won't be as bitter ... and there are so many non-bitter varieties these days."

Jean and her husband Michael run Jones Valley Farm of Spring Green, marketing more than 125 Old World vegetable varieties to customers at the DCFM, including some of the region's best chefs. Want to know more about broccoli Romanesco, cardoons, celeriac, cipollini onions, escarole, fava beans, radicchio or, of course, eggplant? Ask the Martins.

Jean shared this recipe for a streamlined, no-frying-necessary eggplant Parmesan. The original came from 1978's *The Quick and Easy Vegetarian Cookbook*, by Ruth Ann and William Manners. Says Jean, "Since the eggplant comes out as individual rounds or pieces—not as a casserole (as with most recipes)—it can be served as an appetizer as well as a main course."

Cooking note: This recipe calls for prepared tomato sauce. Use Jean's Thick and Easy Tomato Sauce recipe on page 93, or one of the other variations from page 178 or 180.

Oil for greasing baking dish

1 medium-large eggplant or 2 to 3 smaller ones (about 1 pound total)

3/4 cup freshly grated Parmesan, divided

About 3/4 cup homemade or packaged breadcrumbs

About 1/2 cup mayonnaise

About 1 cup Thick and Easy Tomato Sauce

1/2 pound mozzarella, sliced

Heat oven to 375 degrees. Lightly oil a 9-by-13-inch or similarly sized dish. Wash and trim eggplant(s); cut into 1/2-inch-thick slices. Combine 1/4 cup of the Parmesan and all the breadcrumbs in a pie plate or dish.

Working one at a time, spread mayonnaise on both sides of eggplant slices and dip both sides into the breadcrumb mixture, arranging the dipped slices in a single layer on ungreased baking sheet as you go. Bake them 15 minutes, flipping the eggplant midway through the baking time to brown each side. Remove from oven but leave the oven on.

While they're still hot, arrange eggplant slices in the baking dish, slightly overlapping them, if necessary, so they all fit. Spread each slice with a small amount of tomato sauce—you can either put a stripe down the middle of the slices or spread the sauce out on each slice. Top each piece with a slice of mozzarella and sprinkle the remaining 1/2 cup Parmesan over all.

Bake until cheese is bubbly and eggplant is done, 20 to 25 minutes. Serve hot.

Photos Char Thompson

THICK AND EASY TOMATO SAUCE

MAKES 1 1/2 CUPS

2 cans (each 6 ounces) tomato paste

1/2 to 1 teaspoon each dried basil and oregano

1 to 2 teaspoons minced garlic

1 teaspoon brown sugar

1/2 teaspoon salt

1/4 teaspoon pepper

Mix all ingredients in a bowl until well blended. Leftover sauce can used in melts or rice pilaf, on pizza or crostini, and the like.

WISCONSIN SUMMER SWEET CORN SPAGHETTI

2 TO 3 SERVINGS

Italophile and longtime DCFM shopper Mary Karau has been to Italy more than 50 times. She loves to create dishes that are Italian-influenced but also feature local ingredients. Mary came up with this elegant, ingenuously simple dish one August day when sweet corn at the market was peaking. You can find similar recipes online—which proves the old adage that great minds think alike!

This is an ideal entrée for summertime and really showcases Wisconsin sweet corn. "A simple green salad with a lemon vinaigrette would be a good accompaniment," says Mary. "Tomato bruschetta would also complement it."

If you've got a crowd coming over, the recipe can be expanded to serve more people. "The amounts of ingredients are not hard and fast rules. All can be adapted to your taste." Still, Mary's a stickler about a couple of things. She insists on authentic, freshly grated Parmigiano-Reggiano cheese, and fresh herbs. "Dried herbs will not do!"

Mary Karau

Kosher salt

6 to 8 ounces high-quality dried Italian spaghetti (such as De Cecco)

2 to 3 ears of fresh Wisconsin sweet corn

4 tablespoons butter

2 to 3 tablespoons minced shallots

White pepper

Dash of smoked paprika

1 to 2 tablespoons extra-virgin olive oil

Large handful of washed and dried basil leaves, coarsely chopped or cut into thin strips

Small handful of washed and dried Italian parsley, minced or finely chopped

1/4 to 1/2 cup freshly grated Parmigiano-Reggiano

1/2 teaspoon freshly grated lemon zest (optional)

Garnish (optional): 2 to 3 small clumps whole basil, snipped from the top of basil plants

Fill a medium-large pot with filtered or regular water; bring to boil over high heat and add 1 heaping teaspoon salt. Add the spaghetti and return to boil, stirring often at first to prevent clumping. Reduce heat so that water simmers but does not boil over. Cook, following package directions, until pasta is al dente, approximately 8 to 10 minutes.

Meanwhile, place serving bowls in a warm oven or near the boiling pasta water. Cut the kernels off the ears of corn onto a plate. Use the dull side of the knife to scrape remaining bits of corn and the juices from the cobs.

Melt the butter in a large skillet over medium heat. Add shallots and saute until they are soft but not browning. Add the corn kernels and stir until corn is barely cooked, 3 to 5 minutes. (When corn is fragrant, it's probably ready.) Season to taste with salt and white pepper. Stir in a generous dash of smoked paprika.

When the pasta is done, drain it, reserving about 1/2 cup of the water. Stir olive oil into the pasta to keep it from sticking and clumping.

To serve: If needed, add a bit of reserved pasta water to the pasta to moisten it. Mound the pasta in the bowls. Stir chopped basil and parsley into the corn mixture. Spoon corn mixture atop the pasta, dividing evenly between the bowls. Sprinkle 2 or more tablespoons cheese over each bowl. A bit of finely grated lemon zest is totally optional but might be tasty. Garnish each bowl with the top of snipped basil leaf plant.

LESCÓ
(HUNGARIAN TOMATO AND PEPPER STEW)

MAKES ABOUT 8 QUARTS (24 TO 30 LARGE SERVINGS)

Lescó is to Hungary what ratatouille (see page 88) is to France, or shakshuka (see page 206) is to Israel—a humble, one-pot staple that fills the belly and feeds the soul. "Every Hungarian family has a recipe for lescó, a stew of tomatoes, peppers, onions and paprika," says contributor A.B. Orlik, who got the family recipe from her Hungarian grandparents. A.B. is a loyal customer—and occasional helper—at the DCFM's Young Earth Farm, owned by Shirley Young.

"In the early days of the COVID-19 pandemic, when the world shut down, I found myself baking bread and knitting, and when Farmer Shirley brought veggies to Delta Beer Lab on Saturday mornings, my heart (and tummy) turned cartwheels. I bought my first 20-pound box of tomato 'seconds' (these aren't any ordinary seconds; Shirley's standards are high!), picked out some of every pepper she sold, and made an enormous kettle of lescó for the first time since I was a teen. It was gone in days, scooped up by my father, sister, in-laws and friends."

A.B. Orlik

Cooking notes:

A.B.'s stew features an assortment of varieties for each of the main ingredients—peppers, tomatoes and paprika. She has her favorites, but you are welcome to play with them.

For the peppers, she says, "Start with banana peppers as the foundation, but add lots of variety—pimiento, paprika, bell, shishito." (One variety she doesn't use is jalapeño.)

For the dried paprika mixture, she likes 2 to 3 tablespoons sweet Hungarian paprika, about 1/2 tablespoon smoked paprika and about 1 teaspoon of something spicy, like cayenne ("If I'm in the mood for a little kick").

For the tomatoes, aim for an assortment that are vine-ripened, deep-flavored and colorful.

Lescó recipes vary tremendously, in ingredient proportions, the type of fat used, how long to cook it, how spicy or mild to make it, how thick or brothy it ends up being, and what it's served with. You can make a little or a LOT. (It freezes beautifully.) Make it your own.

6 tablespoons lard or coconut, olive or canola oil, plus extra for browning meat

6 to 7 medium-large onions (3 1/2 to 4 pounds), halved lengthwise and sliced into half-rings (8 to 9 cups total)

Kosher salt and pepper

5 to 6 large assorted peppers (see Cooking notes, page 96), cut into thin strips (7 to 8 cups)

3 to 4 tablespoons assorted dried paprika

20 medium-large assorted tomatoes (10 to 12 pounds), cored and cut into small chunks (16 to 20 cups)

2 pounds kielbasa or smoked sausage of choice (optional)

For serving: Boiled or baked potatoes, Hungarian dumplings (small, drop-style dumplings), cooked rice or noodles

A.B. Orlik

Heat a huge, heavy stew pot over a medium-high flame. Add oil or fat of choice and heat it for a moment or two. Once oil is hot, add onions and cook, stirring occasionally, until wilted, 5 to 8 minutes. Sprinkle the onions with 2 to 3 pinches kosher salt about halfway through this initial cooking time.

Add the peppers and cook, stirring occasionally, 3 to 5 minutes. Add assorted paprikas; stir and cook for about 30 seconds, making sure the paprika doesn't scorch. Stir in the tomatoes, all their juices, another big pinch or two of salt and 2 to 3 pinches of pepper. Bring to boil, reduce to a strong simmer, sniff and smile, and then get yourself comfortable. Let the whole thing simmer for an hour or so, stirring occasionally, tasting it for the depth and spiciness of the paprika flavor and adjusting as you like. You'll know it's done when the tomatoes have broken down but the peppers and onions still hold their shape. What you're looking for is a thickish, not-too-brothy consistency. If it's still too brothy for you after an hour or so of cooking, keep going until it's the consistency you like.

If you're adding kielbasa or other sausage, do so at this time: Cut them in half lengthwise and then into 1/2-inch-thick half-moons. Brown them lightly with a little more oil in a frying pan before adding them to the stew. Or if vegetarians or vegans are among the guests, present the meat as a side dish.

Serve the stew over your choice of potatoes, dumplings, rice or noodles.

MULTI-MUSHROOM WRAPS WITH SPRING ALLIUMS, SALAD GREENS AND CHEESE

2 TO 4 SERVINGS

Remember the baby vegetable craze? Those teeny carrots and miniscule beets, the near-invisible-to-the-naked-eye microgreens. At one time they seemed so chic, so very French. But of course good cooks have been making delicious use of immature plants and garden trimmings ever since vegetable seeds first sprouted in Eden.

Green garlic, spring onions and narrow leeks are vibrant-tasting early versions of fully grown alliums, the stage when the narrow shoots have thickened and the bulbs have begun to form but are still fairly small. In this recipe, these tasty shoots are stir-fried with delicate mushrooms grown by Josh and Danielle Clark of Clark Family Gardens, a small urban gourmet mushroom farm. The Clarks pile the mixture inside a springtime quesadilla decked out with melty cheese and fresh greens. It may look like a traditional quesadilla, but according to the recipe tester, it tastes a lot like the spinach *gateau* from Madison's much-missed Ovens of Brittany restaurants.

1 to 3 tablespoons sunflower oil or butter, divided

1 pound mixed mushrooms (such as shiitake, lion's mane, chestnut, oyster), coarsely chopped

1 cup chopped young, narrow-stemmed leeks (white and light green parts only)

1/2 to 3/4 cup sliced spring onion bulbs (or substitute regular onions)

1/4 cup sliced green garlic bulbs or 1 tablespoon minced garlic

Salt and pepper

1/2 cup sour cream

4 flour tortillas (6 to 7 inches diameter)

1/2 cup grated Parmesan

1/2 to 1 cup shredded medium or sharp cheddar

2 to 3 cups salad greens (arugula, baby kale, spinach, etc.)

Heat oven to 350 degrees. In a large skillet, heat 1 to 2 tablespoons of the oil or butter over medium-high flame until very hot. Add mushrooms, leeks, onions and some salt and pepper. Stir and cook the mixture, adjusting the heat if scorching threatens, until mushrooms are tender, 8 to 10 minutes. If pan dries out too much during the cooking, add a little more oil or butter and/or 2 to 4 tablespoons water. Stir in a little more salt and/or pepper, if needed.

Lay tortillas out on a parchment-lined baking sheet. Spread sour cream over half of each tortilla. Sprinkle both types of cheese over the sour cream.

When mushroom-allium mixture is done, divide it evenly over the cheese-covered portion of the tortillas. Bake until cheese is melted, 5 to 7 minutes. Top each tortilla with a fistful of salad greens, fold in half and serve pronto.

Danielle Clark

Danielle Clark

Danielle Clark

Danielle Clark

Char Thomson

RAMPAKOPITA

4 SERVINGS

"I created this recipe to showcase our early season ramps," says Susan Smith, who, with her husband Matt, was a DCFM vendor for 40 years, from 1980 to 2020. "I make this spiral-shaped pastry with a springtime mixture of ramps, spinach, chives and baby greens, like kale. You can even add little nettles. In Greece, it's made with a mix of wild herbs and is called *hortopita*."

Cooking notes:

If you haven't worked with phyllo dough before, you will learn the first time you open a package that it is very delicate and dries out easily. To keep it pliable for assembling the pastry, thaw the boxed dough slowly in the refrigerator ahead of time, and take one of the inner packages out of the fridge about a half-hour before you're ready to use it. Don't open the inner packaging until you're ready to assemble the pastry. (What to do with any leftover sheets? Carefully roll them back up, tape the inner packaging, box up, and re-refrigerate for another use.)

If you have purchased (or harvested) only ramp leaves without the bulbs, you can substitute 1 medium shallot or 1 small onion for the ramp bulb.

For the phyllo coils, use 1 or 2 sheets per coil, depending on your preference for more filling versus more pastry.

1 pound total of ramps, chives, spinach and baby greens (such as kale or nettles), washed well in cold water

Knob of butter, for sauteing

8 ounces feta cheese, crumbled (about 1 1/2 cups total)

1 cup Tzatziki, divided (recipe on page 103)

Salt and pepper

8 or 16 packaged phyllo dough sheets, thawed according to package instructions

4 or so tablespoons butter, melted

1 tablespoon nigella and/or sesame seeds

Cut the white bulbs off the ramps; chop the bulbs and set aside. Heat a large, nonstick skillet over medium flame. Tip in the ramp leaves (but not the chopped bulbs), along with the chives, spinach and greens. Cook until wilted, stirring frequently, about 3 to 4 minutes. (You won't need any oil or water as the greens will steam in their own moisture). Place greens in a colander, let them stand to cool a little, then squeeze out as much liquid as possible. Chop greens and place them in a bowl.

Heat a knob of butter in the skillet. Add the ramp bulbs and cook until softened, about 10 minutes. Tip these into the bowl with the greens, then use a fork to stir in the feta and 1/2 cup of tzatziki until everything is well combined. Taste for seasoning, adding salt and pepper as desired.

Heat the oven to 375 degrees. Brush a round cake pan with a little of the melted butter.

Lay 1 or 2 phyllo sheets on a work surface (one atop the other if there's two) and brush with melted butter. Spread about 2 tablespoons of the filling in a line along the long edge of the phyllo. Roll loosely into a long sausage (don't roll it too tight as this might cause the pastry to split), then coil the pastry into a pinwheel and place in the center of the pan.

Repeat with the rest of the dough and filling, arranging the coils around the central one. Brush the tops with more butter, then sprinkle with the seeds. Bake until golden and crisp, 25 to 30 minutes. Serve hot or warm, with remaining tzatziki.

Deb Shapiro

RAMP IT UP

Technically, ramps are wild leeks, but they don't look or taste like domesticated leeks. You can recognize them by their wide, tapering leaves, wine-red stems and the bulging, creamy white root end. They grow in upper Midwestern woodlands and show up at the market in early May, tied in pretty bunches and ready to get to work in your kitchen.

All three parts of the plant are edible, and they taste like they smell—like garlic, but more herbal and a little sweeter. To store ramps, place them in damp paper towels inside plastic bags; they'll stay fresh for a week or more.

Lightly sauteed or chopped raw ramp leaves can be folded into egg dishes, soft polenta or hot mashed potatoes. The stems and bulbs sweeten when cooked; tuck some inside a grilled cheese sandwich, top a burger with them, and use them to flavor spaghetti sauce or sauteed mushrooms.

TZATZIKI
(CUCUMBER GARLIC YOGURT SAUCE)

MAKES ABOUT 1 CUP

Tzatziki is a zesty cold sauce that's popular in the cuisines of southeast Europe and the Middle East. Enjoy it with lamb or chicken skewers, good bread, rice dishes, raw veggies and yes, even French fries. It's also one of the ingredients in the Rampakopita recipe on page 100.

If you don't have time to strain the yogurt to thicken it, you can use 1/2 cup Greek yogurt instead. And if you like your tzatziki extra-zesty, stir in some chopped fresh mint or dill.

1 cup plain whole milk yogurt

1 small to medium cucumber

1 teaspoon salt

1 tablespoon lemon juice

1 tablespoon olive oil

2 teaspoons minced garlic, mashed with a dash of salt to a paste
(use a fork or the flat of a chef's knife for this)

Place yogurt in a cheesecloth-lined sieve set over a bowl to drain. Refrigerate 3 to 24 hours—the longer it drains the thicker it gets. (Don't toss the liquid; drink it—it's delicious.)

Peel, seed and thinly slice the cucumber, as thin as you can get it. Mix the slices with salt, place in a colander and let stand 30 minutes at room temperature. (This concentrates the cucumber flavor and draws out liquid for a silky texture.) Lightly rinse cucumbers, gather them into both hands and squeeze them to remove liquid. Chop the cuke slices finely. Mix cucumbers, lemon juice, olive oil, garlic paste and strained yogurt.

SPRINGTIME ASPARAGUS LASAGNA

8 TO 10 SERVINGS

"This is a comforting recipe that I look forward to making every year because it offers a great way to use fresh asparagus and welcome spring into your kitchen," writes shopper Emily Rose Haga. She created it by playing with a similar recipe from *Back to the Table: The Reunion of Food and Family*, by Art Smith.

Think of it as a crossover dish, one that pairs the first pencil-thin spears to show up at the market in early spring with the last of the winter-stored potatoes. Or it can shine in early summer, too, when asparagus season is waning and the earliest, tiniest new potatoes appear on the scene. It's rich and subtly flavored, the kind of casserole that's elegant enough for company. To make it extra special, Emily uses fresh pasta (though dried is just fine, too), "and, of course, some good ol' Wisconsin cheeses."

The dish can be prepared and assembled well in advance of baking, even the night before. Let it cool completely, cover and refrigerate it, and then bring it back to room temperature before baking. Leftovers taste wonderful (maybe even better than the first time around).

1 pound thick asparagus spears, halved lengthwise, chopped or left whole

3 medium yellow potatoes (1 to 1 1/4 pounds), peeled or not, cut into 1/4-inch-thick rounds

1 package (12 ounces) fresh lasagna noodles

1 to 2 tablespoons olive oil

Cheese sauce:

6 tablespoons butter

3 tablespoons flour

3 cups whole milk, warmed in the microwave or atop the stove

1 bay leaf

Pinch of nutmeg

Salt and pepper

1/2 cup grated Parmesan

Also:

2 cups (about 8 ounces) grated mild Swiss cheese (or another mild, white cheese), divided

1/2 cup grated Parmesan

2 to 3 tablespoons butter, cut into pieces

Bring a large pot of salted water to a boil over high heat. Add asparagus and cook it until barely tender, about 3 minutes. Use a sieve or tongs to transfer asparagus to a bowl of ice water; swish it around until the asparagus is cool. Drain asparagus and pat dry with towels.

Add potatoes to the still-boiling water and cook until just tender, about 10 minutes. Use a sieve to remove potatoes to a plate, pat them dry and let them cool off. If using fresh pasta, no cooking is necessary before assembling the lasagna. If using dried noodles, cook them in the still-boiling water until barely tender (follow directions on package). Drain noodles and toss gently with a little olive oil to prevent sticking.

Meanwhile, make the cheese sauce: Melt 6 tablespoons butter in a saucepan over medium low heat. Whisk in flour to form a roux. Cook for about 2 minutes, stirring often with a wooden spoon, then whisk in milk, bay leaf and nutmeg, plus salt and pepper to taste. Bring to a simmer. Whisk often and cook until sauce has slightly thickened, about 5 minutes. Remove from heat then whisk in 1/2 cup Parmesan.

Terese Allen

To assemble lasagna: Heat oven to 350 degrees. Lightly butter a deep 9-by-13-inch baking pan or similarly sized baking dish. Spread a thin layer of cheese sauce on the bottom of the pan, then a layer of pasta sheets. Over this, layer 1/3 of the sauce and half each of the asparagus, potatoes and Swiss cheese. Add another layer of pasta, another 1/3 of sauce and the remaining vegetables and Swiss cheese. Top with another layer of pasta, the remaining 1/3 of sauce, and 1/2 cup Parmesan. Dot with butter pieces.

To bake: Place lasagna on middle rack in the oven and bake until golden brown on top and bubbly around the sides, about 30 minutes. (If it's not browning, you can turn on the broiler for the last few minutes if you're using a broiler safe pan.) Let the dish cool for about 10 minutes before serving so the sauce has a chance to set a little before you cut into the lasagna.

THE TRUTH ABOUT LASAGNA

Lasagna is surely one of Italy's greatest gifts to the world. But if you believe that a "classic" lasagna made with wide, flat noodles, meaty ragu and Italian cheeses is the only truly authentic one, then your faith is about to be tested. Italy itself has numerous regional versions that are as beloved as the one that's considered the "real deal" by so many Americans. Ingredients like spinach, eggplant, truffles, radicchio, chicken livers and even hard-boiled eggs play a role in these lasagnas.

What's more, lasagna probably didn't even originate in Italy. Many sources say it came from ancient Greece, where the word *laganon* referred to sheets or strips of pasta, and to a dish composed of layers of pasta and sauce. Conquering Romans brought the idea home, and well, from there on in, the Italians really flew with it. By the Middle Ages, lasagna was showing up in recipe books, and the term referred more to method than to specific ingredients. "The Italians introduced cheese in the recipe around the 1280s," says TheProudItalian.com. "And eggs weren't used in the recipe until the Renaissance. Tomato appeared later in the recipe in the 1880s in Naples."

That's when classic lasagna, as Americans think of it, really began to take shape. To this day, we seem to love that cheesy, meaty, tomato-forward version best. However, we also like to have our lasagna and eat it too—many, many kinds of it. Take a quick spin around the Web and you'll find recipes for Mexican lasagna, vegan lasagna, polenta lasagna, lasagna with goat cheese and pumpkin, chicken bacon alfredo lasagna roll-ups, and even something called "Cinnamon Toast Crunch Dessert Lasagna." Or check out the farmers market, where everything from asparagus to zucchini is available for creating your own.

On page 104, contributor Emily Rose Haga offers her version of lasagna, an elegant dish that celebrates springtime in Wisconsin by showcasing asparagus. "I also like to make another version in the fall and winter using roasted beets," says Emily. "It's a strikingly beautiful recipe to make for someone you love on Valentine's Day!"

Harmony Valley Farm
Certified Organic Produce
Red & Yellow
Pimento Peppers
$3 per pound
rmony Valley Farm
tified Organic Produce
i Sweet

"SOMEWHERE OVER THE RAINBOW" STIR-FRY AND BROWN RICE TOSS

4 TO 6 SERVINGS

Ben Moss and Emery Farah are a couple who regularly turn down other Saturday morning plans in order to be on the Capitol Square. "The DCFM holds a near-and-dear spot in our hearts," says Ben. "We love to walk and talk and smell and taste and enjoy and support the local community."

They also love to eat stir-fries. Emery goes for the vegetables and Ben has a passion for rice and ginger. "This recipe was an experiment to see if we could have it all in one dish," Ben explains. "Turns out we can, and so can you."

New to stir-frying, they learned a good lesson. "When chopping, slicing and ribboning your vegetables, thinner is better," says Ben. "They'll cook faster, and you'll be more likely to achieve that subtle, tasty char that comes with the best stir-fries." Ben and Emery recommend serving their multicolored stir-fry with chicken egg rolls, just the way they do.

1 to 2 cups organic brown rice (make the larger amount if you like a high ratio of rice to veggies)

2 tablespoons sesame oil, plus more as needed

2 cups corn kernels

1 1/2 cups thinly sliced broccolini or broccoli stalks

1 cup Brussels sprout leaves (cut in half, cut out core and pull sprouts apart to separate the leaves)

1 cup ribboned carrots

2 large mushrooms, thinly sliced

2 red radishes, thinly sliced (optional)

1 red Fresno chile pepper (or other medium spicy chile pepper), slivered

1 tablespoon minced fresh ginger

1 tablespoon minced garlic

2 cups cooked, shredded chicken (optional)

5 to 6 tablespoons bottled or homemade kung pao sauce

Garnish (optional): 2 green onions, chopped

Cook the brown rice according to the package directions. When it's done, keep it warm on the back of the stove.

For the stir-fry: Combine corn, broccolini or broccoli, Brussels sprout leaves, carrots, mushrooms, radishes (if using), chile, garlic and ginger in a large bowl; toss well.

Heat sesame oil in a wok or very large, deep, heavy pan over medium-high heat. Add the veggie mixture and cook, stirring often, until veggies are lightly charred and nearly tender, about 15 minutes. Add more sesame oil as the mixture cooks, if necessary. Add rice and chicken (if using) to the pan; cook and stir until everything is well combined and chicken is heated through, 2 to 3 minutes. Add kung pao sauce and stir well until heated through. Serve it up, topped with green onions if you like.

Photos Ben Moss

DOMADA
(WEST AFRICAN PEANUT STEW)

8 TO 12 SERVINGS

"My favorite time of year for the market is September, when tomatoes meet root veggies," writes loyal attendee Meri Tunison. Here she shares a recipe that is the perfect reason why. *Domada* is the national dish of Gambia, a joyous big-pot dish that mingles tomatoes, peppers and cooler weather vegetables with ground peanuts and chicken.

"My parents met and got married while they were in the Peace Corps in The Gambia," says Meri. "They brought a domada recipe back to the States with them. As a child, I was confounded by the idea of hot peanut soup. As an adult, I have come to love this recipe and all other peanut stews."

This hearty, dairy-free stew is customizable to the size of your pot and the kind of vegetables you like best. You can also make it vegan by using a vegetable stock and substituting additional vegetables for the chicken. Serve it with rice or couscous (or not!), and don't skip the toppings—they really highlight the dish.

Meri Tunison

4 to 6 boneless, skinless chicken thighs (or substitute other chicken parts)

4 to 5 cups broth (chicken, beef or vegetable) or water

2 cups diced onions

3 to 4 cups diced tomatoes

1 to 3 small habaneros, jalapeños or other hot peppers, finely chopped (or substitute 1/2 cup diced bell pepper if you want it less spicy)

1 to 2 tablespoons minced garlic

3 tablespoons tomato paste

1/2 teaspoon salt

1/2 teaspoon cayenne pepper

1/4 teaspoon black pepper

3 to 4 cups chopped vegetables such as sweet potatoes, butternut squash, carrots, Brussels sprouts and/or broccoli

1/2 cup natural peanut butter (with no sugar or salt)

Additional peanut butter, tomato paste, broth and seasonings as needed

Lime wedges

Toppings (use any or all): Minced cilantro, diced green onions, chopped peanuts

Place chicken parts in a large pot and cover with the broth or water. Add onions, tomatoes, peppers, garlic, tomato paste, salt, cayenne and black pepper. Bring to a low boil, skimming the froth from the surface as needed. Once at a boil, cook about 5 minutes.

Add vegetables and peanut butter; stir well. Bring to a simmer, adjust heat to maintain a simmer, and cook the stew, stirring occasionally, until the veggies are tender and chicken is cooked through, about 20 to 30 minutes. Remove the chicken, let it cool until you can handle it, then shred it with a fork, or chop or chunk it up with a knife. Stir chicken back into the pot.

Taste the stew and add more peanut butter, tomato paste, broth and seasonings as you like. Simmer for a few more minutes after making final adjustments.

Just before serving, squeeze the juice from a few lime wedges into the stew, then taste and add more as desired. Serve the stew over cooked rice or couscous, topped with cilantro, green onion and/or peanuts.

ROSEMARY-BRINED ROASTED GUINEA FOWL

4 TO 6 SERVINGS

Here's a simple but singular Italianate recipe from Jill Negronida Hampton. Her son and daughter-in-law, Nico Bryant and Melanie Hook, own a pastured poultry operation called Bryant Family Farms, where they raise specialty breeds of chicken, duck, turkey and guinea fowl. The latter is a bird that most DCFM shoppers didn't have access to until the couple started bringing it to market. Endemic to Africa, guinea fowl is a dark-meat bird with lean flesh and a full-bodied flavor that, well, tastes like chicken ... only *better*.

The recipe was adapted from one served at the now-closed Rotisserie Georgette restaurant in New York City. You'll need to plan in advance so that the bird can bathe in a garlic- and rosemary-infused brine for a day or three. For a side dish to go with it, consider the Mushroom and Pea Risotto on page 158.

10 cloves garlic, smashed with the flat of a knife and peeled

1 1/2 cups thinly sliced onions

3 tablespoons chopped fresh rosemary

1 tablespoon chopped fresh thyme

2 tablespoons brown sugar

3 tablespoons salt

1 teaspoon whole black peppercorns

1 guinea fowl (3 1/2 to 4 pounds)

1 to 2 tablespoons softened butter

Combine 10 cups cold water, garlic, onions, rosemary and thyme in a large pot and bring to a simmer; simmer 30 minutes. Remove from heat and stir in another 10 cups cold water, brown sugar, salt and peppercorns. Cool the mixture to room temperature. Completely submerge the guinea fowl in the brine, weighting it down with a small plate if necessary (or transfer bird and brine to a bowl or pot big enough to submerge the bird). Refrigerate it for 1 to 3 days.

Line a rimmed baking sheet with foil and place a rack on top. Remove guinea fowl from fridge, dry it well with paper towels and place it on the rack. Let stand at room temperature 30 to 60 minutes. Meanwhile, heat the oven to 350 degrees.

Roast the bird 40 minutes, then remove it from the oven. Raise oven temperature to 450 degrees. While oven is heating up more, rub the fowl with butter all over. Return bird to oven and continue roasting until it's well browned, approximately 15 minutes. Remove from oven and let stand 7 minutes before serving.

LANDJAEGER JAMBALAYA

8 TO 10 SERVINGS

Jambalaya's appeal comes from a layering of many flavors. The list of ingredients looks long, but it's basically an easy, one-pot creation with no complicated techniques or fancy cooking maneuvers. This take on the down-home Louisiana classic features something unconventional for a traditional jambalaya: landjaeger, the smoked and fermented meat stick snack sold at many a Wisconsin bar, butcher shop and gas station. (You can also get it from Marr Family Farm and other meat stands at the DCFM.) It finishes the Cajun dish the way a jack-matching five card completes a perfect hand of cribbage.

Recipe contributor Jesse Brookstein knows a thing or two—or three thousand—about landjaeger (see page 117). In fact, he wrote the book about it, called *A Perfect Pair: The History of Landjaeger in Green County, Wisconsin*. After one of his research trips for the book, he found himself with an excess of sausage. His mother, a veteran jambalaya maker, came up with the idea to use landjaeger in jambalaya when Jesse sent her a load of them. We've adapted their recipe herein (photo on page 116).

3 tablespoons olive oil, divided

3/4 lb. pound boneless, skinless chicken thighs or breasts, cut into 1-inch pieces (2 1/2 to 3 cups)

2 tablespoons Cajun/Creole seasoning, or more to taste, divided

3 small bell peppers (assorted colors), seeded and diced (about 1 1/4 cups)

1 cup diced white onion

3/4 cup chopped celery

2 tablespoons minced garlic

2 cups chopped fresh tomatoes or 1 can (14 ounces) crushed tomatoes

1 1/2 cups uncooked long-grain white rice

3 1/2 cups chicken broth

1 teaspoon salt, or more to taste (optional, in the event the Cajun/Creole seasoning is heavily salted)

1/2 teaspoon ground black pepper, or more to taste

1 teaspoon dried thyme

1/4 teaspoon cayenne pepper or 1/2 teaspoon dried chile flakes

1 teaspoon Worcestershire sauce, or more to taste

1 bay leaf

1 jalapeño, seeded and finely chopped, or 1 teaspoon bottled hot sauce, or more to taste (optional)

3 to 4 pairs landjaeger sausage, cut into 1/2-inch pieces (2 to 3 cups)

1 cup thinly sliced okra (if frozen, be sure to thaw and drain)

3/4 to 1 pound raw shrimp, peeled and deveined (if frozen, be sure to thaw and drain)

Garnishes: Thinly-sliced green onions and/or chopped fresh parsley

Heat 1 tablespoon of the olive oil in an enameled Dutch oven or large, heavy pot over medium-high heat. Add chicken pieces and 1 tablespoon of the Cajun/Creole seasoning; saute, stirring often, until chicken is cooked through, 5 to 10 minutes. Remove chicken to a bowl, cover and set aside.

Add the remaining 2 tablespoons oil, then the bell peppers, onion, celery and garlic. Cook, stirring occasionally, until the onions are softened and slightly transparent, about 6 minutes.

Stir in tomatoes, rice, chicken broth, salt (if using), black pepper, thyme, cayenne or dried chile flakes, Worcestershire, bay leaf and remaining 1 tablespoon Cajun/Creole seasoning. If desired, add the optional jalapeño or hot sauce.

Bring mixture to a low boil then reduce heat to medium-low and give everything a stir. Cover pot and let mixture cook slowly, stirring occasionally, 10 to 12 minutes. Stir in the landjaeger and continue cooking until the liquid is nearly absorbed and rice is very close to done, another 15 to 20 minutes. While it's cooking, keep an eye on the liquid absorption, adjusting the heat and/or adding a little water so that the rice cooks through and there's enough moisture to help cook the shrimp and okra.

Gently stir in shrimp, okra and reserved chicken. Cover the pot. Cook, stirring occasionally, until the shrimp are pink and cooked through, and okra is softened but still has some give, about 7 to 9 minutes. Remove the bay leaf.

Taste the jambalaya and, if needed, season with additional salt, pepper, Cajun/Creole seasoning and/or hot sauce as you like. To serve, top each plateful with green onions and/or parsley.

This take on the down-home Louisiana classic features something unconventional for a traditional jambalaya: landjaeger.

Recipe on page 114

Jesse Brookstein

LANDJAEGER LOVE

Jesse Brookstein is a walking, talking ode to landjaeger, the old-world sausage that he refers to as a "time-capsule sausage" and a "marvel of a meat stick." In his book, *A Perfect Pair: The History of Landjaeger in Green County, Wisconsin*, Jesse explains that a traditional landjaeger is made of beef, pork or both, and molded into a flattened, rectangular shape, hickory-smoked and typically sold in joined pairs. "The meat is spiced and cured to delay the growth of any spoilage or pathogenic bacteria, and then it's subjected to a fermentation process that drops the pH level to such a spot that those harmful bacteria can no longer grow," he says. "Add some drying to reduce any additional likelihood of bacterial issues, and you have a meat product that can be taken on the road without any refrigeration."

Landjaeger spicing varies from producer to producer, usually featuring some combination of caraway, coriander, black pepper, allspice, celery seed, garlic, cumin and nutmeg. "But even if one producer were able to see another shop's spice bill, there is little chance they could replicate the same exact product due to other factors, such as starter cultures, meat and wood sourcing, in-house practices and general meat shop terroir."

The epicenter of landjaeger culture in Wisconsin is Green County, where Swiss influence is strong. The sausages themselves probably originated in the part of Europe where Germany, Austria, Switzerland and France come together.

Landjaeger, along with jerky and other styles of meat sticks, makes the perfect protein-packed snack for deer hunters, campers and other Wisconsin outdoor enthusiasts—including the folks who make their hungry way around the Capitol Square on Saturday mornings.

Terese Allen

CROATIAN-STYLE STUFFED RED PEPPERS

6 SERVINGS

Recipe contributor Betsy Abramson likes this one-dish preparation because "it's got some zip to it, it's made on the stove top, and it doesn't have any dairy in it." She prefers red, yellow or orange sweet bell peppers over green because of their sweeter flavor. Look for ones that are even-bottomed; they'll stand up better in the dish. Otherwise use a sharp knife to shave a very thin slice off the protruding lobes to even them up. And she encourages cooks to increase the garlic and spices or add cayenne, as you like.

1 1/4 pounds ground turkey or beef

2 to 3 tablespoons olive oil, divided

1 1/2 teaspoons salt, divided

6 medium red, yellow or orange sweet bell peppers

2 cups chopped onions

6 tablespoons chopped fresh parsley

1 to 2 tablespoons minced garlic

2/3 cup cooked white rice (leftover rice works well)

1 tablespoon sweet Hungarian paprika, or more to taste

1 teaspoon ground pepper

1/4 teaspoon ground allspice

1/4 to 1/2 teaspoon cayenne (optional)

1 large egg

2 1/2 to 3 cups homemade or store-bought plain tomato sauce, divided

Garnish (optional): Finely chopped parsley

Heat a large, deep skillet over medium-high flame. If you're using ground turkey or very lean beef, add 1 tablespoon oil and swirl the pan to coat the bottom. Add the meat, sprinkle it with 1 teaspoon salt and brown, breaking it up as it cooks, 6 to 8 minutes. Remove the meat to a medium bowl with a slotted spoon and discard the fat, if there is any.

Cut off the top half-inch of each of the peppers. Cut the flesh away from the pepper stems, discard the stems and finely chop the pieces of flesh. Use a grapefruit spoon or small, sharp knife to scoop out seeds and ribs from the peppers.

Using the pan you used for browning the meat, heat 2 tablespoons olive oil over medium flame. Add the onions, chopped pepper pieces and another 1/2 teaspoon salt; saute them, stirring occasionally, until partially softened, 4 to 5 minutes. Add the parsley and garlic; continue to cook, stirring now and then until the vegetables are tender, another 3 to 5 minutes.

Transfer mixture to the bowl with the meat. Add the rice, paprika, pepper, allspice, cayenne (if using), egg and 1/2 cup of the tomato sauce to the bowl. Stir until well combined.

Fill the pepper cavities with the meat mixture, leaving some space at the top so the stuffing doesn't spill out as it cooks. Pour 1/2 cup of the remaining tomato sauce over the bottom of a large, heavy pot or Dutch oven (enameled cast iron is ideal). Stand the peppers up in a single layer inside it, with just a little space between them. Pour about 1 1/2 cups tomato sauce around the peppers.

Bring sauce to boil over medium-high heat, then reduce heat to low, cover pot and simmer the peppers in the sauce until they're fork-tender, 35 to 45 minutes. Check the dish occasionally as it cooks, basting the peppers with sauce, and/or adding additional sauce if it thickens too much. When the peppers are done, turn off the heat and let them stand, covered, for a few minutes before serving.

To serve, lay the peppers on their side in soup plates or shallow bowls, one per person. Ladle some sauce around and over the peppers. Sprinkle with chopped parsley.

THE STUFF OF CROATIA

Common across southeastern Europe, stuffed peppers are called *punjena paprika* in Croatia and, despite their heartiness, are often eaten in summer, when bell peppers are in season. Ground meat, rice, paprika and parsley are key ingredients in the filling, and cooks serve the peppers atop a simple tomato sauce, usually with mashed potatoes alongside. Beef is traditional, but some cooks use pork or lamb or a combination of two meats. (In the recipe on this and previous page, ground turkey is also an option.) Northern cooks, take note! Since stuffed peppers taste even better the second day (and if one is making them in summer, it can be hot work), some families prepare them in the evening, making enough to last for several days or to freeze for a robust meal during the colder months.

CARNITAS
(AKA MEXICAN PORK CONFIT)

8 OR MORE SERVINGS (ENOUGH TO FILL 12 OR MORE 6-INCH CORN TORTILLAS FOR TACOS)

It's hard to believe that something so easy to make can be so spectacularly delicious. But believe it. Pasture-raised pork and the excellent lard that comes from it are the only ingredients besides salt that you need for the rustic-yet-decadent Mexican specialty called *carnitas*. To make it, simply salt the pork, cover it with melted lard, and braise it until the meat is so tender it shreds easily, and is crusty on the outside and moist on the inside.

This recipe comes from Cliff and Cari Gonyer of Rockwell Ridge Farm in Dodgeville, who sell pasture-raised heritage hog cuts, baked and canned goods, plus eggs and lard at the market. Cari sometimes adds seasoning to the carnitas once it's done, such as garlic, dried chile flakes, cumin or chipotle. "But it is also really good plain," says Cliff. "We serve it with traditional Hispanic toppings, but it is also delicious pan-fried with potatoes, onions and peppers, or added to tomato sauce and served on top of polenta."

And don't stop there—carnitas can be featured in enchiladas, tamales, quesadillas, barbecue pulled pork sandwiches, macaroni and cheese, scrambled eggs and lots more. If you're going to serve it just one way, however, make traditional Tacos de Carnitas (see page 121) and offer a variety of toppings.

1 pork shoulder roast (3 to 4 pounds)

2 pints good-quality lard

About 3 teaspoons kosher salt

Heat oven to 325 degrees. Cut the roast into chunks that are approximately 2-by-3-inches in size. Lightly salt the chunks of pork on all sides (this is an important step; don't skip it). Place them to fit snugly in a 3-quart Dutch oven or other deep, heavy pot with a tight lid.

Melt the lard in a microwave oven or on the stove top. Pour it over the pork, enough to come about 3/4 of the way up the meat (if it doesn't, you may need a smaller pot). Cover the pot tightly and place it in the oven.

Lucinda Ranney

Cook until the pork is very, very tender—done enough to break apart easily with a fork, 2 1/2 to 3 1/2 hours. Some of the meat may also be browned and beginning to crisp up on the outside (this is good).

Use a slotted spoon to remove meat from the pot to a platter. Let stand until it's cool enough to handle, then coarsely shred the meat with your fingers. To serve, reheat it briefly in a pot on the stove, adding a bit of water if you like. (Alternatively, you can reheat it in a heavy skillet with a little of the hot fat, until more of the edges have browned and crisped up.) It's now ready to use in tacos (below) or other dishes.

Later, when the lard has cooled but is still in liquid form, strain it through a fine-mesh sieve into a bowl. When it's fully cool, transfer the lard to a 1-quart jar, secure the lid and store it in the refrigerator. (What about those unctuous tidbits that are left in the strainer? They're the cook's reward.) The lard can be used again (try frying potatoes in it—yummy) and will keep in the fridge for months.

TACOS DE CARNITAS

12 OR MORE SERVINGS

This is fiesta food. This is make-your-own fun. Don't skimp on the toppings with these. Part of the fun is trying new combinations and accidentally-on-purpose overfilling your tortillas.

12 or more good-quality corn tortillas, warmed on a hot griddle or wrapped in foil and heated in the oven

Carnitas (see page 120)

Toppings:

Quick-Pickled Shallots and Radishes (see page 31)

Diced avocados

Chopped green onions

Halved cherry tomatoes

Shredded cheddar or crumbled feta

Sour cream

Refried beans

Salsa Verde (see page 184)

Cilantro leaves

Slap a warm tortilla on a plate, add some carnitas and bring on the suggested toppings. Your choice.

GRILLED PORK TENDERLOIN WITH CILANTRO STEMS AND BLACK PEPPER RUB

4 SERVINGS

Don't throw away cilantro stems—nor the roots either, if you're lucky enough to find them intact when you buy a bunch. They are a key flavor in Thai cooking, with a citrus pungency that softens when they're cooked. DCFM customer Lisa Dussault features them in this adaptation of a recipe from *Hot Sour Salty Sweet: A Culinary Journey Through Southeast Asia*, by Jeffrey Alford and Naomi Duguid. Sticky rice or basmati rice is the ideal accompaniment.

Cooking notes:

You'll need only cilantro roots (if available) and stems for this recipe. You can use the leaves to make Lisa's Cabbage Salad with Soy-Ginger Dressing, page 60, to serve with the pork.

Silver skin is a thin, silvery membrane on a tenderloin. Typically the butcher cuts it off, but if that hasn't been done (or not completely done), you can remove it thusly: Insert a sharp, thin-bladed knife between the silver skin and the flesh, angle it slightly downwards and flat against the silver skin, and then pull up the skin as you lightly scrape the knife against it. Keep doing this until all, or nearly all, of the sliver skin is removed.

Terese Allen

1 small or 1/2 large bunch cilantro, with roots, if available

2 tablespoons bottled fish sauce

1 1/2 to 2 tablespoons coarsely chopped garlic

2 teaspoons coarsely ground black pepper

1 pork tenderloin (about 1 1/2 pounds)

Chile sauce:

1/2 cup sugar

1/2 cup rice or cider vinegar

1 1/2 to 2 teaspoons minced garlic

1 teaspoon dried chile flakes, or more to taste

Dash or two of bottled fish sauce

Salt

To prepare a seasoning paste: Cut the cilantro roots and approximately 1 inch of the stems from the bunch. If there are no roots, trim off a little of the stem ends and discard them, then cut two or so inches of stems from the bottom of the bunch. Clean in cool water. Save the leaves for other uses.

Coarsely chop cilantro roots (if using) and stems. Place in blender or food processor with fish sauce, garlic, black pepper and about 2 tablespoons water. Blend mixture to a smooth paste, scraping down sides of work bowl occasionally and adding a little more water as needed to form a paste.

If it has not already been done, remove silver skin from the pork tenderloin. (See cooking notes on previous page.) Rub seasoning paste over tenderloin and refrigerate for 1 to 2 hours.

About 1/2 hour before serving time, remove tenderloin from refrigerator. Prepare/heat an outdoor grill using wood, charcoal or gas.

While the grill heats, prepare chile sauce: Combine sugar and vinegar in a nonreactive saucepan and bring to a simmer over medium heat. Reduce heat to low and let simmer 5 minutes. Remove from heat and add remaining ingredients. Let mixture cool.

Grill the tenderloin over medium-high heat until the internal temperature reaches 145 degrees, about 20 to 30 minutes. Remove from grill and let rest 8 to 10 minutes. Serve with chile sauce.

CHILE VERDE

4 TO 8 SERVINGS, DEPENDING ON SIZE OF ROAST

"My husband and I love Mexican food," says contributor Lynn Danielson. "We also love pork, but knowing that pigs are very smart animals, we only eat 'happy' pork—that is, local, farm-raised and treated well." She buys much of their meat at the farmers market, particularly in the fall to freeze for winter.

Lynn combines her two loves in this recipe, a merger of several chile verde versions found on the internet that is very flexible, both in ingredient proportions and cooking process, and also in how it's served. You can reduce or ramp up the heat by changing the number of chiles, and use less or more garlic, cumin, salsa and such. You can cook the roast (whole or cut up) in a Dutch oven atop the stove or in a 325-degree oven, or use a slow cooker, like Lynn does. "Some people serve chile verde as a stew, but I like to roll it in flour tortillas," says Lynn. "If you want a stew, add water or chicken stock. Serve it with a salad and perhaps corn bread."

For the chiles, use whatever fresh ones you like and can get—there are many varieties available at the market nowadays. Lynn goes for hot or semi-hot green varieties like jalapeño, Anaheim and poblano, but if you want more kick, go for habanero or Scotch bonnet.

Cooking note: Lynn uses pork shoulder roast in this recipe for its great flavor. But they can be fatty, so she suggests trimming off some or all of the visible fat, and if possible, making the dish a day ahead and refrigerating it overnight. Then you can scrape the solidified fat off the top before reheating and serving.

1 to 2 tablespoons olive or canola oil

1 bone-in or boneless pork shoulder roast (3 to 4 pounds)

2 cups chopped tomatillos

1 1/2 cups chopped onions

2 or more green chiles, seeded and diced

2 tablespoons minced garlic cloves

1 jar (12 ounces) salsa verde, or use the recipe on page 184

1/2 teaspoon cumin

1/2 teaspoon salt

1/4 teaspoon pepper

Pinch of sugar

Heat oil in a deep, heavy skillet or Dutch oven over high heat. Add roast and brown it well on all sides. Combine all remaining ingredients in a slow cooker, then add the browned roast. Cook on low until the meat is falling-apart tender, around 5 hours, stirring occasionally. Remove the roast, cool it enough to handle it, then shred it with a meat fork. Return it to the pot with the sauce and heat through to serve. Alternatively, you can puree the sauce in a blender or food processor, combine it with the shredded meat and heat to serve.

If you want more kick, go for habanero or Scotch bonnet.

MEE'S HMONG MUSTARD GREENS WITH BOILING PORK

6 TO 8 SERVINGS

One of the defining aspects of Hmong cuisine is the role of meat as a component, rather than the centerpiece, of a meal. Another is simplicity, and this recipe is a great example of both. It was shared by vendor Phil Yang and his mother, Mee (see page 145).

"Just mustard greens and pork and salt, boiled," is how Phil describes it. "You can add some pepper if you want, but really, salt is all it needs." And he's right: The greens are herbaceous and peppery, and the pork gives meatiness. It's the kind of one-pot comfort food that makes Phil's face brighten when he talks about it.

Some cooks vary it by stirring in minced fresh ginger or the tender, inner shoots of lemongrass, and diners add heat by enjoying the greens with a fiery fresh chile sauce (see page 127).

One thing Hmong cooks never do, however, is toss the cooking liquid. "It's like a soup," says Phil. "The [watery broth] is an important part." Indeed, lightly salted and enriched, it's hot, healthy and thirst-quenching. To learn more about Hmong cooking, see page 146.

3 to 5 heads mustard greens (about 3 pounds)

1/2 to 1 pound pork belly or baby back pork ribs

2 to 3 teaspoons sea salt or table salt

Rinse the mustard greens thoroughly with cold water. Cut stalks and leaves into 1-inch pieces (you should have about 16 to 20 loosely packed cups total). Rinse the pork belly and cut the meat into 1/2-inch rectangular pieces. (If you're using ribs, cut them into 1-rib pieces.

Bring 2 to 3 quarts water to a boil in a large pot over high heat. Add the meat and let it boil vigorously, skimming off the "bubble" (i.e. froth or foam), until there is no more bubble, about 5 to 10 minutes. Stir in salt, then add all the mustard greens. Bring to a boil and cook, stirring occasionally with a wooden spoon, until mustard greens are wilted and soft, 5 to 7 minutes. Transfer mixture, including the liquid, to a large serving bowl and bring it to the table. Enjoy!

Shutterstock

Terese Allen

SAUCE ON THE SIDE

One must-have dish at nearly every Hmong meal is the spicy pepper condiment called *kua txob ntsw*. Authors Sami Scripter and Sheng Yang have this to say about it in their outstanding book, *Cooking from the Heart: The Hmong Kitchen in America*: "Hmong people simply call this dish 'pepper.' It accompanies almost every meal, and it is never made exactly the same way twice. Good chiles to use are small red and green Thai chiles. Hmong people tend to like it very hot, so the seeds are usually included."

Families place small bowls of the sauce around the table so that everyone has easy access. Below is a recipe for this simple but essential side dish. Garlic, green onion, tender lemongrass leaves, salt and/or sugar may be added to the mix; just mash them in with the chiles and cilantro.

3 to 4 red or green Thai (or other) spicy chiles, chopped

2 tablespoons chopped cilantro

1 tablespoon bottled Thai fish sauce

Juice of 1 lime

Mash the chopped chiles and cilantro with a mortar and pestle until a paste forms. Or mince them with a knife and then mash them with the back of the knife. Stir in fish sauce and lime juice. This makes about 1/3 cup, but it can be multiplied.

Joan Ballweg

GERMAN KOHLRABI IN MEHLSCHWITZE WITH BOILED POTATOES AND SAUSAGES

4 SERVINGS

It had never occurred to Natascha Merten that kohlrabi is considered uncommon in the U.S. until she heard people asking about it at the Square. "Kohlrabi? What is that?"

To Natascha, who became a market regular after moving to Madison from Germany in 2019, kohlrabi dressed with German white gravy (*mehlschwitze*), served with sausages and potatoes, is very traditional comfort food. "My grandma used to make it. My mother used to make it. I like to make it," she says. "We use this kind of gravy in Germany for many Brassica dishes and also for carrots and peas. These veggie sides are always combined with potatoes (boiled is the classic) and combined with meat—sausage, meatballs, meatloaf, pan-fried pork tenderloin, schnitzel, etc."

Natascha's family cooks the dish *Pi mal Daumen*, a phrase that is akin to "rule of thumb" in English and refers to following an approximate method for doing something. "It will be slightly different every time," says Natascha, "but you are pretty confident it will turn out right in the end."

In that same spirit, she gives options for varying the meal: "Sauteed mushrooms are another side dish that fits. You can sprinkle parsley over the potatoes. You can use the fat from the sausage pan as a topping for the potatoes."

But Pi mal Daumen does have its limits. If you want to keep things authentic, says Natascha, "just don't add hot sauce or BBQ sauce."

"Kohlrabi?
What is that?"

1 1/3 pounds baby potatoes (about 4 cups golf-ball size potatoes)

Salt

2 medium kohlrabies, peeled and cut into sticks about 1 1/4 inches long and 1/4 inch wide (2 to 3 cups)

Small amount oil for frying sausages

4 links Polish sausage, kielbasa or andouille, or plant-based sausage

2 tablespoons butter

2 tablespoons plus 1 teaspoon flour, divided

1/2 cup milk or oat milk

Pepper

1/4 teaspoon ground nutmeg, or more to taste

Garnish (optional): 1 to 2 tablespoons chopped fresh parsley or dill

Scrub and rinse the potatoes, place in a pot and cover with cold water. Add 1 teaspoon salt to pot. Bring to boil over high heat, then reduce heat to maintain a simmer. Cook potatoes until softened, approximately 20 minutes.

While the potatoes are cooking, place the kohlrabi in a separate pot, cover with cold water and add 1 teaspoon salt to pot. Bring to a boil over high heat.

As the water heats, start frying the sausages by heating a little oil in a skillet over medium-low heat. Add the sausages and cook them slowly, turning often, until browned and fully done, about 15 to 18 minutes. (If you're using plant-based sausage, see package instructions for cooking time.)

When the kohlrabi has come to a boil, reduce heat to maintain a slight simmer and cook kohlrabi until they are just tender, 3 to 5 minutes. They should have a similar texture as you would like broccoli stems to feel when they are cooked. Remove 1/2 cup of the cooking liquid, combine it with the milk in a small bowl and set this mixture aside. Drain the kohlrabi through a strainer.

Now make the gravy using the same pot used for cooking the kohlrabi: Place pot over medium-low heat, add butter and let it melt. Sprinkle in half of the flour while continuously stirring with a wooden spoon. Stir in 1 tablespoon of the milk-water mixture. Keep stirring as you sprinkle in the rest of the flour. Switch to a whisk, and whisk the mixture while slowly adding more liquid until it has the desired thickness of a gravy. Add salt, pepper and nutmeg to taste. Reduce heat to low and let gravy bubble slowly for a few more minutes, stirring often. Gently stir the cooked kohlrabi into the gravy, and keep it warm for serving.

Check on the potatoes: Use a knife or fork to test if they have softened and are done, then drain off the water, leaving potatoes in the pot. (You can peel the potatoes at this point, if desired, but fresh ones from the farmers market do not need that.) Cover pot with a kitchen towel and then with the pot lid and set aside. This step keeps potatoes warm and prevents sogginess.

When everything is ready, serve the potatoes—sprinkled with parsley or dill—with the sausages and kohlrabi in mehlschwitze. You can use the fat from the sausage pan as a topping for the potatoes, if desired.

Natascha Merten

Lucinda Ranney

KIMCHI FRIED RICE

2 SERVINGS

Contributor Holly De Ruyter found this recipe online from the New York Times. She likes it because it comes together quickly and is a great way to use up items in the fridge, such as rice left over from takeout. "Sometimes I throw in whatever market veggies need to be used up. It helps stop food waste and makes the dish even easier to throw together."

The main feature of the dish is kimchi, a spicy fermented cabbage (or radishes or other vegetables) that is big in Korean cuisine. Holly favors the local stuff made by Kingfisher Farm and usually serves some of it on the side with, as well as in, the fried rice. Another highlight is butter, a more unexpected addition to fried rice than kimchi, but one that mellows and flavors the sauce beautifully.

3 tablespoons butter, divided

1/2 cup finely diced onion

1 cup kimchi, chopped

3 tablespoons kimchi juice

1 cup diced ham, smoked turkey or chicken, or baked tofu

2 cups cooked rice, fully cooled or cold

2 teaspoons soy sauce

1 teaspoon toasted sesame oil

2 eggs

Salt and pepper

Garnishes (optional): Toasted sesame seeds, sliced green onions, crumbled nori, additional kimchi

Terese Allen

Melt 2 tablespoons of the butter in a large skillet (nonstick or cast iron is best) over medium heat. Add onions and cook, stirring often, until partially tender, about 3 minutes. Add kimchi, kimchi juice and protein of choice. Cook, stirring often, until liquid is somewhat reduced, about 3 to 5 minutes. Add rice, breaking it up and stirring it into the mixture. Continue cooking until rice is hot and has absorbed the rest of the sauce, 4 to 5 minutes. Add soy sauce and sesame oil; stir well, then taste and add more soy sauce, sesame oil or kimchi juice as you like. Keep cooking to allow rice to brown a little—or a lot, if you like a crunchy texture.

Heat the remaining 1 tablespoon of butter in a separate, nonstick pan. Fry the eggs in it to your liking, adding a little salt and pepper to taste.

To serve, portion the fried rice onto two plates, add a fried egg to each and garnish with any or all of the optional toppings.

Terese Allen

HAIL TO KIMCHI

Kimchi is hot in more ways than one. The funky, salty and decidedly spicy Korean pickle has hit the big time—not only can you find it at farmers markets now, but it's also getting featured in everything from grain bowls and tofu bowls to scrambled eggs and pasta. Even the juice gets put to good use in dips, dressings and sauces. But here's one that surprised us: Kimchi burritos. Local carpenter and fan of Kingfisher Farm Jim Hansen shares his kind-of-a-recipe below. It makes a great one-handed breakfast (or lunch or dinner) entrée.

Place a clean dishtowel on a plate. Place an 8-inch flour tortilla on the dishtowel. Put cold, cooked pinto beans and rice along with cheese on the tortilla. Add salt and pepper to taste. Microwave for two minutes. Put on an ample amount of cold Kingfisher Farm Kimchi. Fold over the sides of the tortilla first and then roll. Roll the finished burrito in the towel and make a mad dash for work. When you're hungry, unroll the towel and enjoy. You'll want the towel because it's gonna drip.

Like meat? Put some in. Prefer colby to aged sharp cheddar? Your choice. Black beans instead of pintos? Go for it. Want a vegan burrito? Use the Kingfisher Farm Vegetarian Kimchi. Have celiac? Use corn tortillas and make tacos.

If one burrito is not enough, I recommend having two.

BISON OSSO BUCO

6 TO 8 HEARTY SERVINGS

Market members since 1991, Leroy and Cindy Fricke of Cherokee Bison Farms raise purebred, grassfed American bison and offer a full range of fresh cuts as well as bison summer sausage, braunschweiger, snack sticks and jerky. (Another native food they offer is pure maple syrup—the sap harvested from a 650-tree sugarbush, collected by hand and cooked in flat-pans over wood fires.)

For her variation of a classic Milanese *osso buco*, Cindy subs in meaty bison bones for veal shanks, and braises them into tender submission in a tightly lidded heavy pot. We've tailored things a bit to work with either fresh or processed ingredients.

You can use dried herbs, canned tomatoes and store-bought stock here, and you'll have a perfectly delightful meal. But make it with fresh-clipped herbs, garden-grown tomatoes and a gelatinous homemade stock, and you will witness some joyous licking of plates. As a side dish, any of the following would work: polenta, risotto, mashed potatoes, pasta or pureed winter squash. (Truth is, the sauce is so exceptional, it could make a door knob taste great.)

Salt and pepper

4 pounds meaty bison bones, cut from the shank into 2-inch thick pieces

1/4 cup flour

4 tablespoons butter or olive oil

1 1/2 cups chopped onions

1 cup thickly sliced carrots or carrot chunks

1 to 2 tablespoons minced garlic

1 to 1 1/2 cups diced tomatoes (fresh, or frozen and thawed)

1 tablespoon chopped fresh rosemary or 1 1/2 teaspoons dried

2 teaspoons minced fresh thyme or 1 teaspoon dried thyme leaves

1 to 2 tablespoons chopped fresh lovage or 2 teaspoons dried (optional)

1 teaspoon fennel seeds

1 cup dry white wine

1 1/2 cups tomato sauce (preferably homemade)

2 to 3 cups rich beef or chicken stock

Heat oven to 325 degrees. Generously salt and pepper the meaty bones on both sides, then dredge them in the flour and shake off the excess. Heat butter or oil in a large, heavy Dutch oven (enameled cast-iron is ideal) over medium-high heat for a minute or two. When the fat is good and hot, add meat and brown on both sides, about 4 minutes per side. (If necessary, work in batches to avoid crowding the meat.) Reduce heat to medium; remove meat to a platter and set it aside.

Add onions, carrots and garlic to pot; cook, stirring occasionally with a wooden spoon, about 3 minutes. Stir in tomatoes, rosemary, thyme, lovage (if using) and fennel seeds. Add 1/2 teaspoon pepper and, if you're using unsalted homemade stock, add 1 teaspoon salt. (If you're using canned stock, don't add any salt at this point.) Cook the mixture, stirring occasionally, 2 to 3 minutes. Add wine, bring to strong simmer and cook another 2 to 3 minutes, stirring often to scrape up all the tasty bits on the bottom. Return meat to the pot, raise heat to high and add tomato sauce and just enough stock to barely cover the meat. Bring to simmer, stirring the liquid to combine everything well.

Cover pot tightly and transfer it to the oven. Cook until bison is profoundly tender, 1 1/2 to 2 1/2 hours. You'll know it's done when you insert a meat fork into it and the meat slides off the fork.

Remove pot from oven, transfer bison to a pan or platter and place it near the back of the stove to keep warm. Let the sauce in the pot settle for a few minutes, then skim off and discard any fat that has risen to the surface. Raise heat to medium-high and lightly boil the liquid, stirring frequently, until it thickens desirably. Once it hits the thickness you like, season sauce to taste with salt and pepper. At this point, you can get fancy and strain the sauce so that it becomes elegantly smooth. Or you can go rustic and serve it as is. Either way, return the meat to the sauce and gently heat through before serving.

"...the sauce is so exceptional, it could make a door knob taste great."

FASOLIA
(LIBYAN WHITE BEAN AND LAMB STEW)

6 GENEROUS SERVINGS

Run, don't walk, to the market, pick up the main ingredients below, and then hit the stove to make this fabulous stew. It comes from Zainab Hassen, a shopper who loves the large variety of food available at market. "I also use my trips there as an excuse to walk up State Street, grab a cup of coffee and pastry and enjoy the beauty of downtown Madison," says Zainab. Here's her story about this recipe:

> Fasolia *is the Arabic word for beans (it means white kidney beans in Libya), and the term is often used in reference to a classic bean stew that is made across North Africa and the Middle East. This recipe, taught to me by my mother, comes from Libya and is an everyday dish composed of white beans cooked in a tomato- and lamb-based stew. It's a nostalgic dish for most Libyans. I'm an American Libyan [and] it reminds me of back home, my childhood and my beautiful mother.*
>
> *Cannellini (aka white kidney) beans are traditionally used; however, other white beans can be substituted. It's also traditional to make it with lamb, but the dish can be made with beef, chicken, or can even be vegan. The meat is not the star of the show—the white beans are. The meat is only complementing the overall flavor and providing an extra source of protein. That's the philosophy of the Mediterranean diet. Don't be shy with the olive oil, though; the dish tastes much better when it's on the oily side!*
>
> *It is traditionally eaten with your hands and with bread—pita, French or any bread that will readily soak up the stew. However, it's perfectly fine to use a spoon ... and the stew can also be eaten with rice instead of bread.*

Zainab Hassen

Note that fasolia should be a little spicy. Zainab likes it at about the heat of a jalapeño, but it's up to you. "Traditionally, Libyan food is spicy enough, but it should not be painful."

1 1/2 to 2 cups dried cannellini, borlotti or other dried white beans

4 tablespoons extra-virgin olive oil

1 1/2 to 2 cups chopped white onions

2 pounds lamb shoulder meat, cut into large chunks

2 tablespoons tomato paste

1 1/2 to 2 tablespoons minced garlic

Salt

1 1/2 teaspoons ground cumin

1 teaspoon cayenne, or more to taste

1 teaspoon turmeric

1/4 cup chopped cilantro

Hot pita bread or cooked white rice

Garnishes (optional): Chopped green onions and cilantro leaves

The night before you plan to serve the stew, place beans in a large bowl or pot, cover with water by 2 to 3 inches and let them soak overnight.

The next day, cook the stew: Heat the olive oil in a large, heavy-bottomed pot over medium-high heat. Add the onions and cook them, stirring often, until translucent, 7 to 10 minutes. Add lamb pieces to the pot; turn them as they cook to brown them on all sides. Stir in tomato paste and garlic; cook 1 to 2 minutes. Add a pinch of salt (more will be added later). Now season the lamb with cumin, cayenne and turmeric. Cook, stirring once or twice, until fragrant, 1-to 2 minutes. Add enough water to nearly cover the lamb. Bring to a simmer, reduce heat, stir pot, partially cover it, and slowly simmer the lamb until it is fork-tender. This will take about 1 1/2 to 2 1/2 hours.

While the stew is cooking, drain the soaked beans, place them in a pot, cover with cold water, add a large pinch of salt and place over high heat. Bring water to a low boil, reduce heat and let beans simmer slowly until barely tender. This may take up to an hour or more. Do not overcook the beans; they will cook more when added to the stew. Once they're cooked—that is tender all the way through, but not split open—turn off heat and set pot aside until the lamb is done.

To finish the stew: Reheat the beans and then drain them, retaining a cup or two of the liquid, which has starch in it now and can be used to thicken the stew. Add beans and chopped cilantro to the lamb. Simmer the stew uncovered until it reduces a little and reaches the desired thickness; it should be on the thicker side, like a bisque. If desired, you may add some of the reserved bean liquid to help thicken the stew. Season with salt to taste, and add additional cayenne, if desired.

Serve hot with pita bread or rice. Serve with green onions and cilantro leaves, if you choose, to sprinkle over the stew.

WHOLE-SPICE GOAT CURRY

6 TO 8 SERVINGS

It may seem incongruous at first glance, but a booth that sells soap and goat meat makes perfect sense at the farmers market. Lori Hoyt, owner of Soap of the Earth, raises herbs, flowers, bees and goats for the production of fragrant goat milk soaps, hand-stirred in small batches on an 1872 farmstead and sold at market. She also peddles frozen goat meat cuts, which star in this knock-out curry. (Don't worry, there's no soap in it.)

Lori modified a recipe from *The Essential Andhra Cookbook, with Hyderabadi and Telengana Specialties*, by Bilkees Latif, which features foods from India's southeastern coastal areas. During her first encounter with food of the Andhra region, she learned that spices are used whole and are meant to be navigated around in the mouth. Lori says, "I bit into a cardamom pod with my first bite, a star anise pod, my second. Then curry leaves, coriander seeds and fenugreek! The flavor intensity was completely thrilling. It was different than other Indian cuisine I'd had before."

Lori serves the dish with basmati rice, cucumbers, green onions, hard-boiled eggs, yogurt and a squeeze of fresh key lime or lemon—a stupendous meal. If there's any left over, the sauce can be turned into a delicious cross-cultural breakfast dish, one that mimics Israeli shakshuka. "Simply portion it out into individual ramekins, make an indent in the centers and drop an egg into each hole. Pop them in the oven until the egg is cooked and serve. It's wonderful."
(See more about shakshuka on page 206.)

1/2 cup store-bought or homemade clarified butter or ghee (see page 140), or olive oil

3 medium long-shaped white potatoes, quartered lengthwise

2 to 2 1/2 cups thinly sliced yellow onions

5 cardamom pods

5 whole cloves

1-inch piece of cinnamon stick

1 to 2 whole star anise pods

1 teaspoon each coriander seeds, fennel seeds and cumin seeds

1 to 2 tablespoons minced fresh ginger

1 to 2 tablespoons minced garlic

1/2 teaspoon each turmeric and dried chile flakes

2 pounds goat meat, cut into chunks

1 teaspoon caraway seeds

3 cups rough-chopped fresh or home-canned tomatoes,
or 1 can (28 ounces) whole tomatoes with juice, roughly chopped

1 cup each chopped fresh mint and cilantro

For serving:

Cooked basmati rice

Finely chopped cucumber

Finely chopped green onions

Hard-cooked egg, halved or quartered

Plain yogurt

Fresh key lime or lemon wedges

Heat ghee or olive oil in a large, deep skillet or wide, heavy stewpot over medium-high flame. When fat is hot, add the potatoes and brown them well, turning occasionally, 5 to 8 minutes. Transfer potatoes to a bowl. Add onions to skillet and cook until golden, stirring often, 8 to 10 minutes. Stir in cardamom, cloves, cinnamon stick, star anise, coriander, fennel, cumin, ginger, garlic, turmeric and chile flakes. Cook 1 to 2 minutes until fragrant, stirring once or twice. Add goat meat and saute, stirring occasionally, until browned, about 10 minutes. (You may need to adjust the heat up to brown it or turn it down if it seems like it will scorch). Now stir in the caraway and enough tomatoes to barely cover the meat. Bring to a simmer, reduce heat to low, cover pot, and slowly simmer the stew until meat is very tender, 2 to 3 hours.

Stir in the reserved potatoes, mint and cilantro, and continue to simmer until potatoes are tender, 20 to 25 minutes.

Serve hot with basmati rice, and pass around bowlfuls of chopped cucumber, green onions, hard-cooked eggs, yogurt, and lime or lemon wedges for guests to add as they like.

If there's any left over, the sauce can be turned into a delicious cross-cultural breakfast dish, one that mimics Israeli shakshuka.

BUTTER UP: GHEE VS. CLARIFIED BUTTER

Is there a difference between clarified butter and ghee? Clarified butter is butter from which the milk solids and water have been removed, so that it can be cooked at a high temperature without burning. It's not difficult to make: Bring cut-up butter to a simmer over a low flame (don't stir it or let it boil). Skim off the white milk solids that rise to the top, and then spoon out the clear gold butterfat, leaving the watery liquid at the bottom in the pan. The rich-tasting, clear fat is now clarified and ready to use. The leftover milk solids and water can be used to flavor rice pilaf, soups and other preparations.

Ghee is one type of clarified butter, typically used in Indian cooking. The process is similar, but instead of removing the milk solids right after they've risen to the surface, you let them cook until they've browned a little (no burning, please!) and then strain them out. The resulting butterfat tastes wonderfully toasty.

Lucinda Ranney

Recipe on page 142
Sarah Brooks

LITA'S ZUCCHINI, CUCUMBER AND GROUND MEAT

ABOUT 8 SERVINGS

Adriana Mateus came to Madison from Colombia at age 11 with vivid memories of food adventures she and her grandmother had shared together back home. "Walking to local markets near our home in Bogotá to purchase fresh produce, bread and other ingredients was something my grandma—who I called Lita—and I did together nearly daily while I was growing up," writes Adriana. "While I knew nothing about cooking then, I recall Lita's careful examination of potatoes, tomatoes and any other items that would go into our basket of goodies to be turned into delicious meals.

"After we moved to Madison, our family attended the Dane County Farmers' Market from time to time, [but it wasn't] until I returned to Madison in 2005 that I became a much more attentive and frequent visitor. Having spent a few years in Chicago being exposed to amazing French, Spanish, Italian, Mexican and other cuisines, I had both a new appreciation and desire to learn about cooking. Indeed, Julia Child's advice that there's no substitute for fresh, high-quality meats and veggies is something that has been a constant in my life since a very young age."

Adriana contributed this special recipe, one her Lita and mom introduced to the family. "As far as I know, this is not a traditional Colombian recipe, for I have not seen it in any book or eaten it anywhere but at our home." But a key part of the dish is *hogao*, the Colombian creole sauce made with tomatoes and green onions. Like the *sofrito* of Spain, hogao is used as a fragrant, deeply flavored base or as a dipping sauce for many dishes.

Lita's creation is hearty and filling; nevertheless, it has a lightness that gravy-minded Americans might like for a change. Serve it for lunch or dinner. (Photo on page 141.) *¡Buen provecho!*

Cooking note: If you have one, use a mandoline to slice the zucchini and cucumbers. If not, use a sharp knife and slice them as thinly as you can.

For the meat:

1 1/2 pounds lean ground beef or pork, or a combination of the two

1 to 2 tablespoons minced garlic

1 teaspoon dried thyme or 3 to 4 fresh thyme sprigs

2 bay leaves

1 generous teaspoon salt

1/2 teaspoon pepper

1 tablespoon olive oil

For the hogao:

2 tablespoons olive oil

1/2 cup finely chopped green onion, using primarily the white part

1 to 2 teaspoons minced garlic

About 7 plum tomatoes, peeled, seeded and chopped, or pureed (1 1/2 to 2 cups total)

1 teaspoon ground cumin, or more to taste

1/4 teaspoon each salt and pepper

Also:

5 small yellow and green zucchini, each 5- to 6-inches long, sliced very thinly (3 1/2 cups)

2 cucumbers, each 6- to 7-inches long, very thinly sliced (2 to 3 cups)

Cooked white rice

For the meat: If time allows, season the meat the night before for the best flavoring; otherwise, do this anytime within a few hours to a few minutes prior to cooking (it will still taste delicious). Season it with the garlic, thyme and bay leaves, salt and pepper.

Heat the oil in a large, deep skillet over medium-high flame. When it's hot, add the meat and break it apart with a spoon; let it cook thoroughly without drying out. Throughout the cooking, remove some of the extra fat in the pan.

For the hogao: Heat the olive oil in a saucepan over medium-low flame and then add the green onions and garlic. Cook, stirring occasionally, until they've taken on some color but aren't browned, 5 to 7 minutes. Add the tomatoes, salt and pepper, cover pan and cook 10 to 20 minutes. The sauce should have a lot of *sabor* (taste), so cook it until it has deep flavor, adding more salt, pepper and cumin as needed. Its texture should be similar to a thin salsa.

Putting it together: Add the hogao, zucchini and cucumbers to the cooked meat. You will see a lot of veggies, but remember that they will release water and blend in nicely with the meat and the hogao. Cover and cook over medium-low heat until the veggies are tender but still keep their shape, 10 to 15 minutes. Adjust the seasonings once again, as needed.

Serve with white rice and enjoy an exquisite meal with loved ones and perhaps with some wonderful traditional Colombian Vallenato music.

Lucinda Ranney

MEE'S GREEN BEANS WITH GROUND BEEF, GARLIC AND THAI PEPPERS

6 SERVINGS

DCFM member Phil Yang shared this Hmong specialty, a recipe from his mother, Mee, who spent three decades growing and selling market vegetables while raising a family of eight children. In 2018, Phil began offering sweet pastries from a second family business at the market, named Yummee Treats. Appropriately, the bakery is named after Phil's mom, combined with a word that connotes the pleasure of good food.

Ingredient amounts in this recipe are purposefully approximate. Hmong families have a penchant for the intensity of garlic and hot peppers, for example, so are more likely to use the larger amounts given here. But feel free to adjust for personal taste. Serve the dish with jasmine rice and a bowl of fiery chile sauce (see page 127), for the heat-seekers at your table.

1 to 2 1/2 pounds fresh green beans

3 to 4 tablespoons vegetable oil

1 pound ground beef

3 to 5 cloves garlic, sliced or finely chopped

3 to 5 dried Thai hot peppers

1 to 2 teaspoons of sea salt or regular salt

Rinse green beans well, trim off the stem ends, and cut them in half.

Heat oil in an extra-large, heavy skillet over high flame. Add the ground beef, garlic and Thai peppers; cook, stirring often, until meat is beginning to brown, 2 to 3 minutes. Add the green beans and 1 teaspoon salt; toss and cook them until browned in spots and as tender as you prefer, 5 to 10 minutes. Add additional salt to taste. Transfer mixture to a large platter and serve.

HMONG COOKING

Based in an ancient farming culture, Hmong cuisine features a huge variety of health-giving, seasonal vegetables and simple, vibrantly flavored preparations. Hmong immigrants to Wisconsin began to influence farmers markets—and the diversity of foodways in the region—not long after they came to the state as refugees of the Vietnam War.

At the market, vendors supply such familiar Asian favorites as cilantro, cucumbers and hot peppers, but they offer more exotic fare, too, from curly pea shoot tendrils and gnarly-looking bitter melons to lemongrass and foot-long asparagus beans. Frugality is almost an art form in Hmong culture. Gardeners harvest pea shoots and the prunings from squash plants, then sell them by the bunch to be used fresh in salads, stir-fried as a side dish or simmered in nourishing broths.

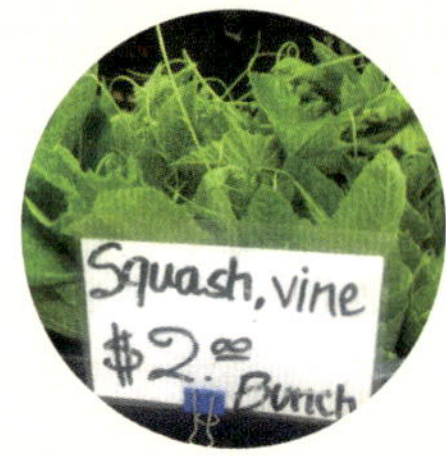

Terese Allen

Hmong diners eat a great deal of cabbage, lettuce, cucumber, squash and many varieties of nutritious greens. Most meals include fresh herbs, especially mint, basil and cilantro. Green onions and garlic flavor many dishes. For a fragrant, lemony touch in soups or boiled chicken, cooks add slender green stalks of lemongrass. The very young shoots are tender enough to be finely minced and added to stir-fries, but mature lemongrass, cut into large pieces and used more like bay leaves, is typically removed before serving.

Lucinda Ranney

Another favorite crop is Thai eggplant, an egg-sized, grassy-green orb that has a burst of white veins on the base and a stem that extends claw-like around the top half of the fruit. Small ones can be eaten raw with salt and pepper for a crunchy and pleasantly bitter snack, and any size can be cooked with such additions as chicken, lemongrass, garlic or basil.

Terese Allen

Like many Southeast Asians, Hmong people appreciate bitter-tasting foods, and favor bitter melon, a summer specialty. It looks like a pale-green, deeply wrinkled cucumber but is used like zucchini, especially in stir-fries with chicken. The strong flavor can be tamed for less adventurous palates by salting it as you would eggplant. Your first taste—pre-salted or not—can be a shock, but like hot peppers or quinine water, bitter melons have a distinctive flavor that could become a passion.

Shutterstock

Rice is fundamental to the Hmong diet. Some people eat long-grain white rice at every meal, whereas mellow-flavored sticky or "sweet" rice is an occasional treat. Spicy chiles are another essential ingredient. At mealtime, families place bowls of very spicy dipping sauce strategically around the table. The sauce is often made with Thai chiles—a mere inch or two in length and almost overpowering to the Western tongue. They're no problem for Hmong diners, however; in fact, some will tell you that store-bought hot peppers are not hot enough! You've got to get them fresh from the grower.

For an excellent cookbook about both traditional and Hmong American cooking, seek out *Cooking from the Heart: The Hmong Kitchen in America*, by Sami Scripter and Sheng Yang.

Lois Bergerson

Rachel Murphy

COWBOY BEANS

6 LARGE OR 12 SMALLER SERVINGS

The name of this dish says a lot. Think: Cattle drive, hungry men and hearty food. Dried beans were easy to transport and cook on the cattle trail, and beef was a staple throughout the West. Modern renditions of the dish—and there are a million of them—usually feature ground beef, but it makes sense that camp cooks of the old days would have carried dried or cured meat with them to fortify their bean pots.

So it's not as much of a stretch as you might think that this recipe for cowboy beans features beef sticks. It comes from Murph's, a family farm whose market stand is often mobbed by shoppers trying to decide which of their eight varieties to munch on. (Pet lovers buy their dog bones, too, and people with a sweet tooth also pick up Murph's maple syrup.)

"Living on the farm, we grew up eating cowboy beans," says Rachel Murphy, who owns Murph's. "But now it's fun to change up the recipe and add our own ingredients." She gives a Tex-Mex twist to this version with their jalapeño-flavored beef sticks. It makes a very meaty bean dish, robust enough to be a main course, but you can also use half the meat for a more classic side dish.

2 tablespoons bacon fat or vegetable oil

1 to 1 1/2 cups finely chopped yellow onions

2 to 3 teaspoons minced garlic

9 Murph's Jalapeño Beef Sticks (12 to 14 ounces total), sliced 1/4-inch thick to yield about 3 cups

1 can (28 ounces) Bush's Original Baked Beans, with liquid from can

1 can (15 ounces) cannellini beans, drained

1 can (15 ounces) black beans, drained

1 cup ketchup

1/4 cup pure maple syrup

1/4 cup apple cider vinegar

2 tablespoons yellow mustard

Heat the fat or oil in a large, heavy pot over medium flame. Add the onions and cook them, stirring occasionally, until tender, about 10 minutes. Add garlic and cook another few minutes.

Stir in all remaining ingredients and bring to a simmer. Reduce heat to low, partially cover the pot and cook beans 30 to 45 minutes, stirring often.

The beans are good to eat at this point, but they improve in flavor if you turn off the heat and let them cool completely, stirring occasionally. Reheat to serve or refrigerate until needed (they even freeze well).

KOREAN-STYLE BBQ BEEF RIBS

4 TO 6 SERVINGS

Jim and Rebecca Goodman helped broaden our understanding of what it means to "eat sustainably" when they began bringing pasture-raised, certified organic beef to a farmers market that was largely about vegetables. Jim's grandfather established Northwood Farm in 1889 and it remained in the family for two more generations. "As a small farm striving to stay in business, we diversified by raising our dairy steers. We were honored to vend that beef at the DCFM for 18 years," says Jim.

"During those years, our customers would request cuts we knew nothing about. Korean-style ribs were a good addition to cuts we offered—they cook faster and really soak up a marinade since they are a thinner cut. The cut was easy for our processor to do, and it changed our short ribs from a surplus item into one that was always in short supply."

The cut, also called "flanken," are short ribs that are cut across the bone, creating a wide strip of meat that's about 1/2 to 1 inch thick, with flat bones along one edge. You can still ask for these at the market; although the Goodmans are retired, other beef vendors will special-order the cut from their butchers. (And they're often available from whole-animal retail butchers.) Flank or skirt steak may be substituted for Korean-style ribs, though they won't have the extra savor of a bone-in cut.

Flanken ribs have an enjoyable chewiness when marinated and grilled quickly the Korean way. "I added stove-top cooking to make them tender, and I finish them on the grill for a smoked flavor and crusty exterior," says Jim. "Cabbage slaw goes well with them." We recommend the Cabbage Salad with Soy-Ginger Dressing on page 60, along with fluffy white rice to complete the meal.

1/2 cup soy sauce

1/4 to 1/3 cup finely chopped green onions

2 to 3 tablespoons finely chopped garlic

2 tablespoons toasted sesame oil

1 teaspoon sesame seeds

1 tablespoon brown sugar, or more if you like it sweet

1 teaspoon black pepper

2 1/2 to 3 pounds Korean-style (flanken-cut) beef short ribs

2 tablespoons canola or other oil, plus a little more for oiling the grill grate

To make the marinade, combine everything but the ribs and canola oil in a dish or pan that will hold the ribs just a little snugly. Whisk well to combine. Add the ribs and move them around in the marinade until well coated. Cover and refrigerate 2 to 8 hours, turning them occasionally.

Remove ribs from marinade (save that marinade!) and dry them well with paper towels. Heat 2 tablespoons of oil in a large, heavy skillet over a fairly high flame. Add the ribs and brown them quickly on both sides. Add the marinade and, if needed, enough water to mostly cover the ribs. Bring to simmer, reduce heat to low, cover skillet and let the meat simmer very gently until mostly done, 20 to 30 minutes, depending on the thickness of the ribs.

While ribs are simmering, heat an outdoor gas grill or prepare coals for a wood or charcoal fire, with a goal of medium-high heat for finishing the ribs.

Remove ribs to a sheet pan, letting excess marinade drain off; pour marinade into a small bowl. Lightly oil the grate. Flash-grill the ribs, turning and basting them often with some of the marinade, about 3 to 5 minutes per side. Frequent turning is necessary so that the ribs char but don't burn. (If the flames get too feisty, spritz them lightly with water to calm them down.) Serve the ribs hot, sprinkled with green onions, if desired.

Deb Shapiro

STEAK AU POIVRE

2 SERVINGS

This recipe for a French bistro classic comes from Jeannie Bull of Bull's Ranch, a small 100% grass-fed beef farm located in the state's picturesque Driftless region. Jeannie adds meaty cremini mushrooms to the usual recipe for *steak au poivre*—if, that is, one can call pepper-encrusted filet mignon topped with cognac cream sauce "usual."

She has plenty of advice for ensuring success: "Any steak may be used, but a tenderloin works wonderfully with the rich sauce. Cast iron is best for getting the right sear on the steak; a nonstick pan will not give the same results.

"Be careful not to overcook the meat, especially since you're using grass-fed beef, which is leaner than corn-fed and cooks best at a slightly lower temperature. What you're aiming for is a good sear for the first minute on each side, then a slightly lower temperature to finish it. If you're not sure of the cooking time, use an instant-read thermometer. And remember, your steak will continue to cook when it's taken out of the pan, so it's important to undershoot the temperature by a few degrees to be sure you don't overcook it."

To mimic the easy elegance of the steak, Jeannie serves it with grilled asparagus, rustic French baguette and a crisp green salad with heirloom tomatoes.

2 grass-fed beef tenderloin medallions, each 6 to 8 ounces and
1 1/2 to 2 inches thick, at room temperature

1 to 2 teaspoons whole black peppercorns, coarsely crushed in a pepper mill, mortar and pestle or with a rolling pin

1 1/2 to 2 tablespoons clarified unsalted butter, divided (see page 140)

1 1/2 to 2 tablespoons grapeseed, canola or peanut oil, divided

2 tablespoons minced shallots

8 to 10 cremini mushrooms, gently wiped clean and then sliced

Kosher salt and ground black pepper

3 tablespoons cognac or good brandy

1/4 cup heavy cream

"Cast iron is best for getting the right sear on the steak."

This recipe moves quickly, so before you begin to cook, have all ingredients prepared as described above. Bring the steaks to room temperature and read through the instructions below.

Press the coarsely crushed pepper onto both top and bottom of the steaks. Place an empty plate near the back of the stove so it warms up as you cook.

Heat a medium cast iron skillet over medium flame for 1 to 2 minutes. When the pan is hot, add about 1 tablespoon each of clarified butter and oil, and swirl pan to coat the bottom. Add the shallots and saute them, stirring occasionally, until they are starting to soften, 2 to 4 minutes.

Add the mushrooms and continue sauteing until lightly browned and tender, 5 to 8 minutes. Season lightly with kosher salt and ground pepper. Use a slotted spoon to remove mushroom mixture to a bowl; set aside.

Turn heat up to high and add more clarified butter and oil, if needed—just enough to coat the bottom of the skillet. Sprinkle the steaks with kosher salt on both sides. Gently place them in the hot fat and cook 1 minute, then turn heat down slightly and continue cooking for about another 3 minutes, for a total of 4 to 5 minutes on the first side, depending on the thickness of the steaks.

Turn heat back up to high and, using tongs (not a fork, which lets the lovely juices run out), turn the steaks over and cook the other side for 1 minute on high heat and then 2 to 3 minutes on a slightly lower flame, for a total of 3 to 4 minutes on the second side. If you're using an instant-read thermometer, the meat should be at about 125 degrees (rare) at this point.

Reduce heat to medium, transfer steaks to the warm plate and cover tightly with foil. Let the steaks rest for 5 to 10 minutes at the back of the stove while you prepare the sauce. While it rests, the meat will continue to cook a little and will reach medium-rare. (If you prefer a medium-cooked steak, add a scant minute of cooking on each side to the above instructions.)

Add cognac to the hot pan, taking care not to splash the liquid. (Some cooks ignite the cognac—this is fancy and a bit dangerous and fun, but not necessary.) Cook the cognac about 2 minutes, stirring up any browned bits on the bottom of the skillet. Add the cream and cook, stirring constantly, until the mixture thickens slightly, another 2 to 3 minutes. While it's simmering, you can add any juices that have accumulated beneath the meat to the pan.

Add the mushroom mixture to the sauce and heat through. Uncover and plate up the tenderloins. Spoon the thickened sauce onto the plates and over the steak medallions. Serve immediately.

Lucinda Ranney

Lucinda Ranney

SIDE DISHES

French Tarragon
-Bright Anise flavor
-A real Sensation
Parsley
-versatile
-a kitchen staple
Wormwood
-silvery foliage
-large perennial 5ft
-Was used to make Absinthe
Sorrel
-bright & lemony
-piquant
Sage
-slight peppery flavor
-excellent cooked with
tomatoes & pork

BOK CHOY WITH GARLIC SAUCE

4 SERVINGS

Holly De Ruyter is a filmmaker known for her portrayals of Dairyland people, places and traditions. In recent years, her documentary, *Old Fashioned: The Story of the Wisconsin Supper Club*, helped inspire a resurgence of interest in this cherished and idiosyncratic institution. Wisconsinites love their supper clubs, Holly has said, because "they want to go somewhere where they know the food is better and know where their food is coming from. It's such a trend now, but that's what supper clubs always were."

Farmers markets and supper clubs are alike that way. "People get so excited, and they are so proud of [their] place in our culture," says Holly. "I like to remind them that it is important we patronize and support [them]. It's part of who we are."

Wisconsin born and bred herself, Holly cherishes supper-club-style old fashioneds and fish fries, as well as cheese curds and corn on the cob from the farmers market. But she enjoys flavors from outside the state, too. She first tried this bok choy side dish after watching a video of it being made at CJEatsRecipes.com. CJ is a third-generation Chinese American home cook whom Holly follows on Instagram. "I do a lot of Asian cooking at home, so this is right up my alley," she says.

"Vegetable sides are often overlooked and can sometimes be bland. This is great with grilled meats. But beware! It will steal the show from the main dish." Truth be told, it's tasty enough to *be* the main dish, served with white or brown rice and sprinkled with roasted peanuts.

Salt

1 1/2 pounds bok choy

2 tablespoons bottled oyster sauce

2 tablespoons soy sauce

1 tablespoon *Shaoxing* wine or dry white wine

1 teaspoon honey or maple syrup

1/4 teaspoon ground white pepper

1 1/2 teaspoons sesame oil

1/2 cup chicken stock

1 1/2 teaspoons cornstarch

1 teaspoon avocado or canola oil

3 to 4 tablespoons minced garlic

Bring 3 or so quarts of water and about a tablespoon of salt to boil. Meanwhile, rinse bok choy well. For small heads, trim off the bottom edge without cutting through the stems, then slice the heads in half lengthwise. For larger boy choy, cut off the bottom inch or so of the head and then thickly slice or chop the stalks and leaves.

Make the sauce: Combine oyster sauce, soy sauce, wine, honey or maple syrup, pepper, sesame oil, chicken stock and cornstarch in a bowl, stirring well to combine.

When the water boils, add the bok choy and blanch until leaves are bright green, 1 to 2 minutes. Drain well and spread out on a cloth towel to dry while you finish the dish.

To finish the dish: Heat oil in a large, heavy skillet over a medium-high flame for a minute or so. Add the garlic and stir-fry until fragrant and beginning to soften, about 20 to 30 seconds, then add the sauce. Simmer, stirring often, until sauce is thickened, 2 to 3 minutes. Arrange the bok choy on a platter. Spoon sauce over bok choy and serve up.

Terese Allen

"Vegetable sides are often overlooked and can sometimes be bland. This is great with grilled meats. But beware! It will steal the show from the main dish."

Lucinda Ranney

MUSHROOM AND PEA RISOTTO

6 SERVINGS

Creamy, comforting risotto is a traditional dish of northern Italian cooking. We knew we had to have a version of it for this book because risotto is so well-suited to seasonal variations of ingredients from the market. This one leans toward springtime and is from Rebecca Buehl-Reichert, who self-describes as "a proud DCFM customer." (She certainly is a loyal one—she's been attending for 25-plus years, even during COVID-19 spikes, when her shopping continued uninterrupted via online ordering and open-trunk pickup markets.)

Choose one or a combination of farmers market mushrooms for this risotto: chanterelles, creminis, buttons, portobellos, oysters, shiitakes, morels, maitakes, hen-of-the-woods, etc. And do like Rebecca does—serve it with your favorite chilled white wine. "Hopefully the same one you used in the dish," as Rebecca says. (Photo on page 160.)

6 to 7 cups mushroom, chicken or vegetable stock

Salt and freshly ground pepper

2 tablespoons extra-virgin olive oil

1/2 cup finely chopped sweet white onion or 1/4 cup minced shallots

3/4 to 1 pound mixture of wild or cultivated mushrooms,
sliced or torn into smaller pieces if they're large

2 teaspoons minced garlic

2 teaspoons fresh thyme leaves or chopped sage

1 1/2 cups arborio rice

1/2 cup dry white wine

1 to 1 1/2 cups blanched fresh peas (or thawed frozen peas)

2 tablespoons chopped flat-leaf parsley

1/2 cup grated Parmesan cheese or other hard grating cheese, plus more for garnish

1 to 2 tablespoons butter (optional)

Garnish (optional): Fresh thyme or parsley sprigs

Bring stock to a simmer in a saucepan. Season it with salt and pepper. (If using canned stock, take care not to oversalt it—you may need none at all.) Keep it hot over the lowest flame.

Heat olive oil in a large, deep, heavy skillet or saucepan over medium flame. Add onions or shallots and cook, stirring often until mostly tender, 3 to 5 minutes. Increase heat to medium-high and add mushrooms; cook, stirring often, 3 to 4 minutes. Reduce heat again to medium, stir in garlic and thyme or sage and season with salt and pepper. Continue cooking until mushrooms are about three-quarters tender, 2 to 4 minutes longer.

Add rice to the mushrooms and stir for 2 to 3 minutes so that every grain gets coated with oil in the mixture. Add white wine and cook, stirring nearly continuously, until most of it has been absorbed by the rice.

Stir in enough stock to barely cover the rice. When it begins to simmer, adjust the heat down to maintain a low simmer. Cook the rice, stirring nearly continuously, until stock is mostly absorbed. Add another ladleful or two of stock and continue cooking, stirring often and adding more stock when liquid is low, until rice is *al dente*; that is to say, tender but with a little resistance at the center. Cooking the rice should take a total of 25 to 35 minutes (taste it for texture—you may or may not use all the stock). Adjust the heat if it's cooking too quickly or slowly.

To finish, add the peas and a bit more stock, if needed, stirring over the heat until peas are hot, 2 to 4 minutes. Stir in parsley and 1/2 cup of the Parmesan. To enrich the dish, stir in 1 to 2 tablespoons butter, if desired. Taste and adjust seasonings, if needed.

To serve, scoop the risotto into soup plates (wide, shallow bowls) and finish each serving with a sprinkle of Parmesan cheese and sprig of thyme or parsley.

Terese Allen

Recipe on page 158

Rebecca Buehl-Reichert

SUSAN'S ERIC'S MOM'S ASIAN ASPARAGUS

8 SERVINGS

Think of this Asian-themed asparagus as the eligible bachelor side dish of spring and early summer potlucks. Contributor Susan Smith, who operated the Blue Valley Gardens market stand with her husband, Matt, for 40 years, got the recipe originally from chef Eric Rupert when he was at L'Etoile restaurant in Madison during the 1990s.

"It was his mom's recipe and I have tweaked it a bit," writes Susan. (We in turn have tweaked it a bit more.) "I make a double batch of the dressing and keep it in the fridge during asparagus season. It's my go-to party dish—people get mad if I don't bring it. And they always want the recipe."

This makes more dressing than you'll need, so you can halve the amount you make or use the leftovers on other blanched veggies—baby carrots, snap peas, fava beans, etc.

2 pounds fresh asparagus, trimmed to even lengths

1 cup good-quality light soy sauce

1/4 cup brown sugar

2 tablespoons rice vinegar

1 tablespoon sunflower (or other mild-flavored) oil

1 to 2 tablespoons sesame oil

Garnish (optional): 1 tablespoon toasted sesame seeds

Prepare a bowl of ice water and a pot of boiling salted water. Plunge asparagus into boiling water until it turns bright green, 1 to 2 minutes, depending on their thickness. (If the spears vary in thickness, immerse the fattest ones first and wait a moment before adding the thinner ones.) Drain and immediately place in the ice water to "shock" the asparagus—that is to say, to stop the cooking and cool it rapidly. Once the asparagus is cool, drain it again and let it dry on towels. Arrange the spears on a pretty plate, cover and refrigerate.

For the dressing, mix soy sauce, brown sugar, rice vinegar and the oils. Just before serving, pour about 1/2 cup dressing over asparagus and garnish with sesame seeds.

GOLDEN ASPARAGUS WITH POTATOES, CHERRY TOMATOES AND NEPALI SPICES

4 SERVINGS

Laura Schmidli is a huge fan of Himal Chuli, a long-lived and much-beloved Nepalese restaurant on State Street in Madison. Laura writes, "This is my attempt to recreate some of the lovely seasonal flavors I've experienced there over the years." She likes combining potatoes and cherry tomatoes with the asparagus but encourages cooks to try other veggies, too. And she notes, "This recipe works best with a cast iron skillet and metal spatula to scrape up the crispy potato bits." Serve it with rice or naan.

Cooking note: *Asafoetida* is onion-tasting seasoning used in Indian cooking, often to flavor dried beans and other vegetarian dishes. Look for it in Indian food stores or, for a substitution, use twice the amount of onion powder.

1 tablespoon ghee (see page 140) or olive oil

3 cloves minced garlic (about 1 heaping tablespoon)

1/2 teaspoon mustard seeds

1/2 teaspoon coriander seeds

1/2 teaspoon asafoetida (or substitute 1 teaspoon onion powder)

1/4 teaspoon turmeric

3 small waxy-type potatoes, peeled and sliced thinly into bite-sized pieces

Handful of cherry tomatoes, halved

1 bunch asparagus, trimmed and cut into 2-inch pieces (or leave them whole, if you prefer)

Salt (optional)

Garnishes (optional): 1 to 2 tablespoons lemon juice and a handful of cilantro leaves

---

"This recipe works best with a cast iron skillet and metal spatula to scrape up the crispy potato bits."

Add the ghee or oil to a large, deep skillet (preferably cast iron) set over medium-high heat. After a moment or two, when the fat is hot, add garlic and spices; stir them until fragrant, about 30 seconds. Be careful not to burn them. Add sliced potatoes and toss to coat. Let them cook and brown for about 5 minutes, stirring once or twice.

Add tomatoes and about 1/4 cup of water. Cover skillet with a tight lid and reduce heat to low.

After about 10 minutes, test the potatoes with a fork. Once the potatoes are almost completely tender, raise the heat back up, add the asparagus and salt to taste (if you like). Saute another 3 minutes, depending on thickness of asparagus, taking care not to overcook it. If you like crispy potatoes, raise the heat to high, and keep your eye on things.

Remove mixture from heat, place on a platter and sprinkle with lemon juice and cilantro. Serve immediately.

BRAISED BABY FENNEL

4 SERVINGS

Pernod is an anise liqueur that harmonizes with similar-tasting fennel in this French-leaning recipe. If the taste of "licorice" with vegetables sounds odd to you, please give it a try anyway. Fresh fennel and Pernod both soften in tone when cooked, and when it comes to flavor, well, you know you can trust the French. (But if you still can't wrap your head around the idea, then substitute dry white wine, or even water for the Pernod.)

The contributor is Rainer Dronzek, who noticed the "beautiful baby fennel" that was available one spring at the Jones Valley Farm stand, where artisanal varietals are plentiful. He adapted this from one in the Williams-Sonoma Savoring Series, *Savoring Provence*, by Diane Holuigue (Time-Life Books, 2002).

"We moved to Verona, Wisconsin in 2016, and one of the reasons was the Dane County Farmers' Market," writes Rainer. "When we lived in the Chicago area, we'd come up occasionally to shop the market. It's one of, if not the best, in the country."

8 to 10 baby fennel bulbs

2 tablespoons olive oil

1 medium yellow onion, sliced (about 1 cup)

1 medium carrot, peeled and sliced on the diagonal (about 3/4 cup)

2 garlic cloves, thinly sliced

1/2 bell pepper, any color, cored and sliced (about 1/2 cup)

1 large tomato, diced—Purple Cherokee is recommended (about 1 1/4 cups)

1 tablespoon tomato paste

2 tablespoons Pernod liqueur

1/4 teaspoon salt

1/4 teaspoon freshly ground pepper

1/4 lemon (optional)

Garnish (optional): Thinly sliced green onions

"The DCFM is one of, if not the best, in the country."

Trim off the stalks of the fennel, leaving about a 2-inch stalk on the bulb. Remove any damaged areas and trim off the bottom. (Save the stalks and fronds for another use or compost them.)

Heat the olive oil over medium heat in a large saucepan with a tight-fitting lid. Add the onion, carrot and garlic, and saute for about 2 minutes, stirring occasionally. Add the fennel bulbs, bell pepper, diced tomatoes, tomato paste, Pernod, salt and pepper. Stir well, cover pan, reduce the heat to low and simmer until the fennel is tender, 10 to 15 minutes. Adjust salt and pepper to taste. Squeeze the lemon over the mixture, if desired, and serve it immediately with a sprinkling of green onions over the top.

BRAZILIAN-STYLE GARLIC COLLARDS (OR MUSTARD GREENS)

4 SERVINGS

"Eat more collards!" says Philip Kauth, executive director of REAP Food Group, the ground-breaking food and sustainability organization based in Madison. Phil learned about the method for these feisty, garden-fresh greens from his colleague Ira Wallace at Southern Exposure Seed Exchange. "Typically, Southern-style collards are cooked for hours with bacon or ham. When my friend tried this recipe [while] on a trip to Jamaica, she was blown away by how simple and easy it was. I've made it for my daughters, and they eat all of it before I get any."

Flash-cooked collard greens are a popular side dish in Brazil. Phil has also used mustard greens in the recipe, and notes that their spiciness diminishes during the cooking.

1 1/2 pounds collard greens, stems removed

1 medium red onion

3 to 6 garlic cloves

1 tablespoon olive oil

Salt and pepper

Splash of soy sauce or red wine vinegar (optional)

Working in batches, stack the collard greens, roll them up and then cut crosswise into thin strips. Finely chop the onion, and mince or coarsely chop the garlic. (Use as much or as little of the garlic as you like.)

Heat the olive oil in a large, wide skillet over medium-high flame. Add the onion and garlic and cook, tossing them often, for just 1 to 2 minutes. Add the collards, toss and cook until they're tender and bright, 3 to 4 minutes. Add salt and pepper to taste, and a splash of soy sauce or red wine vinegar, if desired. Serve up.

Joan Ballweg

Produce of USA
Produce of USA

SWEET AND SOUR KALE

4 TO 6 SERVINGS

The popularity of sweet-sour flavor combinations is widespread—think Cantonese sweet-and-sour pork, German sauerbraten, French *duck à l'orange*, and the barbecue sauces of the U.S. South.

Or how about Minnesota? Market-lover Catherine Jagoe first tasted this sweet-sour kale while at a vacation cabin in northern Minnesota. A friend had found the recipe online, and together they tweaked the dressing to get its sweet-tart balance to their liking, then subbed dried blueberries for cranberries because that's what they had on hand. We also added a few options to the recipe for you to have your own fun with it. (This is how good cooking happens, people.)

"It's become one of my favorite kale recipes. I could eat it every day," says Catherine. "It has a little sauce of its own, so it's good with rice, or as an accompaniment to fish, baked tofu, chicken, steak, pork chops, etc."

2 tablespoons olive oil

1/2 cup diced onion

1 tablespoon minced garlic

1 tablespoon Dijon mustard

1 to 2 teaspoons sugar, maple sugar or honey

1 tablespoon cider vinegar

1 1/2 cups chicken or vegetable broth (plus an additional 1/2 cup if serving the kale with rice)

1 large or 2 medium-small bunches curly or lacinato kale, stemmed, leaves torn into largish pieces (4 to 5 packed cups)

1/4 cup dried blueberries or dried cranberries

Salt and pepper

Cooked white rice (optional)

Garnish (optional): 1/4 to 1/2 cup chopped walnuts, toasted in dry skillet for a few minutes

Heat olive oil in a large, wide saucepan or deep skillet over medium heat. Add onion and garlic and cook, stirring, until tender and translucent, 5 to 7 minutes. Meanwhile, mix mustard, sugar, vinegar and broth in a bowl.

Stir broth mixture into onions; bring to a boil. Stir in kale, cover the pan tightly and cook until kale is wilted, 2 to 4 minutes. Stir in dried fruit and continue to boil uncovered, stirring now and then, 10 to 15 minutes. Liquid should be reduced by half or more. If you're serving this as a side dish, the liquid should be almost gone but pan should not be dry. If it will be served with rice, it can be left saucier. Season with salt and pepper to taste. Serve with rice or by itself. Sprinkle toasted walnuts over the top.

Ruth Bronston

FLYTE'S FRIED GREEN TOMATOES (WITH VARIATIONS)

4 TO 6 SERVINGS

Here's a recipe that comes in handy for Northern growers in autumn, when they harvest unripe tomatoes just before the first frost hits. At the farmers market, we can also get green tomatoes at other times of the year from vendors like Flyte Family Farm, a 3,000-acre vegetable farm in central Wisconsin whose considerable output includes both hydroponic and grown-in-dirt tomatoes.

Carolyn Flyte's version of fried green tomatoes has a mid-20th century feel to it, and when she suggested that Parmesan could be added to the batter as a variation, it got us thinking about other ways to change up the recipe. For Italian flair, for example, you could add fresh basil *and* Parmesan to the batter and serve marinara sauce alongside the tomatoes. Or try these similar-but-different variations on a theme: curry seasoning, yogurt (Indian); Cajun seasoning, remoulade sauce (Cajun-Creole); chopped cilantro, fresh salsa (Mexican).

4 medium to large green (underripe) tomatoes

2 eggs

1/2 cup milk

2 cups Bisquick™ Original Pancake & Baking Mix

2 tablespoons seasoned salt

1/2 to 1 cup butter-flavored shortening, or more as needed

Dill dip or ranch dressing

Slice the tomatoes 1/2-inch thick. Place a large sheet of wax paper near them to hold the tomatoes after they've been battered. Whisk eggs and milk until smooth in a medium bowl. Measure Bisquick™ into a deep plate or pie pan; whisk in the seasoned salt until well combined.

Working one at a time, dip the tomatoes into egg-milk mixture with one hand, place in Bisquick™ dish, and using the other hand, coat each slice well with the baking mix. (Using the same hand for the wet dip and the other for the dry dip each time prevents "gloppy" hands.) Place the dipped slices on the wax paper as you finish each one.

Heat oven to 200 degrees. Set a large, heavy skillet (cast iron is perfect) over a medium-high flame. Add enough shortening to reach a 1/2-inch depth when melted. Keep heating it until a small sprinkle of Bisquick™ begins to sizzle upon contact with the fat, about 5 to 7 minutes.

Fry the tomatoes in batches of 4 or 5, depending on the size of your skillet. Do not crowd the tomatoes—they should not touch each other—and adjust the heat if necessary to prevent overbrowning. When the tomatoes are nicely browned on the first side, flip and fry them on the other side. Each batch will take about 3 to 4 minutes total, and you may need to add and heat a little more shortening to keep the level up for each batch. When each batch is done, drain the tomatoes on paper towels and keep them in the warm oven until they're all done. Or serve them as you finish each batch. Serve with dill dip or ranch dressing.

Rachel Figueora

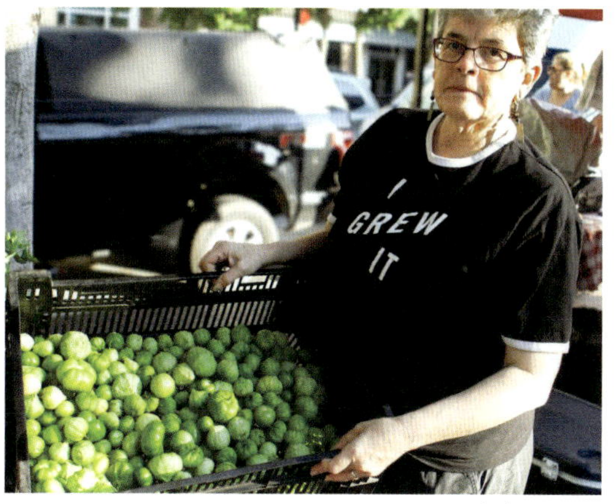

Lucinda Ranney

SURPRISE, SURPRISE

Fried green tomatoes are a traditional Southern dish ... right? Wrong, as we found out when this recipe came in and a little food-history research ensued. Turns out they don't have Southern origins at all. According to Charleston food historian Robert E. Moss, author of *The Fried Green Tomato Swindle and Other Southern Culinary Adventures*, fried green tomatoes "entered the American culinary scene in the Northeast and Midwest, perhaps with a link to Jewish immigrants, and from there moved onto the menu of the home-economics school of cooking teachers who flourished in the United States in the early-to-mid 20th century."

In fact, it wasn't until the movie "Fried Green Tomatoes" (based on a novel by Fannie Flagg) became a hit in the early 1990s that the dish gained popularity in the South ... so much so that it ultimately became a beloved Southern foodway after all.

MAINE SOUR MUSTARD PICKLES

MAKES 1 QUART

This is an old Down East recipe, courtesy of Maine-raised Lisa Dussault, a faithful shopper at the DCFM. "We used to eat them on Saturday nights with baked beans," says Lisa. "My grandfather had a large garden, so we had seasonal produce and canned goods all year."

If you like your pickles sour but with a little bit of sweet, these are the ones for you. The recipe, which can be multiplied as many times as you like, calls for pickling cucumbers, which are shorter and thinner-skinned and have denser flesh than slicing or "salad" cucumbers. Not only do they fit into pickle jars better, they also maintain their crispness better during the pickling process, too.

Pickling cucumbers, enough to fill a quart jar

1 to 2 tablespoons sugar

1 tablespoon dry mustard

1 tablespoon pickling salt

1 to 1 1/2 cups cider vinegar

Wash the cucumbers, scrubbing them lightly to remove dirt, if necessary. If they're very small, you can use them whole. If they're larger, halve or quarter them lengthwise. Pack the cukes snugly but not tightly into a quart jar. Add sugar (1 tablespoon for less sweet, 2 tablespoons for more), dry mustard and salt. Fill the jar three-quarters full with cider vinegar, then add enough water to cover the cucumbers. Screw the lid onto the jar tightly, then shake well to distribute the flavorings and liquid. Leave the jar on a counter for 24 hours, shaking jar occasionally.

After 24 hours, place jar in the refrigerator and allow cucumbers to develop flavor over time. They'll taste good right after the first day, but the flavor will be more developed and balanced after about three days. They will keep in the fridge for a few months.

THREE WAYS WITH GREEN BEANS OR ASPARAGUS

2 LARGE OR 3 TO 4 SMALLER SERVINGS

Megan Bjella is a dedicated gardener-farmer and creative home cook who expanded this recipe from one she found in *Christopher Kimball's Milk Street* magazine. Here, Megan guides us through two slightly different cooking processes—one for asparagus, the other for green beans—and offers three internationally focused sauces to complement them. Take your pick.

Steaming before searing works well for green beans, but for asparagus, not so much. Steaming asparagus first makes it a bit too soggy, interfering with the browning that gives the dish depth. Thus, for asparagus, the method is reversed. For both vegetables, though, as Megan points out, "You reap the benefits of both brightening and softening from steaming, and caramelization and char from searing." And whether you choose beans or asparagus, the dish is tasty as a side, or it becomes a full meal when served over rice alongside a protein of choice.

Green Beans

2 tablespoons grapeseed or other neutral, high-heat oil

1 pound green beans, stem ends trimmed off

Salt

Sauce of choice from pages 176–177

Place green beans in a saucepan; add enough water to come about 1/2 inch up the sides of pan. Cover pan, bring water to boil and steam the beans until they're bright green but still crisp, 1 to 2 minutes. Drain the beans. (If you're not going to finish cooking them right away, immerse them in ice water to stop the cooking and set the color, then drain and dry them off.)

Heat oil in large skillet over high heat until the oil is shimmering. Add the beans, sprinkle with salt, and cook, tossing once or twice until lightly charred, about 2 to 4 minutes. Don't fuss with them too much or they won't develop any color. Top beans or toss with sauce of choice.

Asparagus

2 tablespoons grapeseed or other neutral, high-heat oil

1 pound medium-thick asparagus, tough ends trimmed off

Salt

Sauce of choice from page 176–177

Heat oil in a large skillet over medium-high flame. Add asparagus and let it sear, turning it only once or twice, 2 to 4 minutes. Then add 1/2 cup water to the pan, cover tightly and let it steam until spears are barely tender, 1 to 2 minutes. Drain. Top asparagus or toss with sauce of choice.

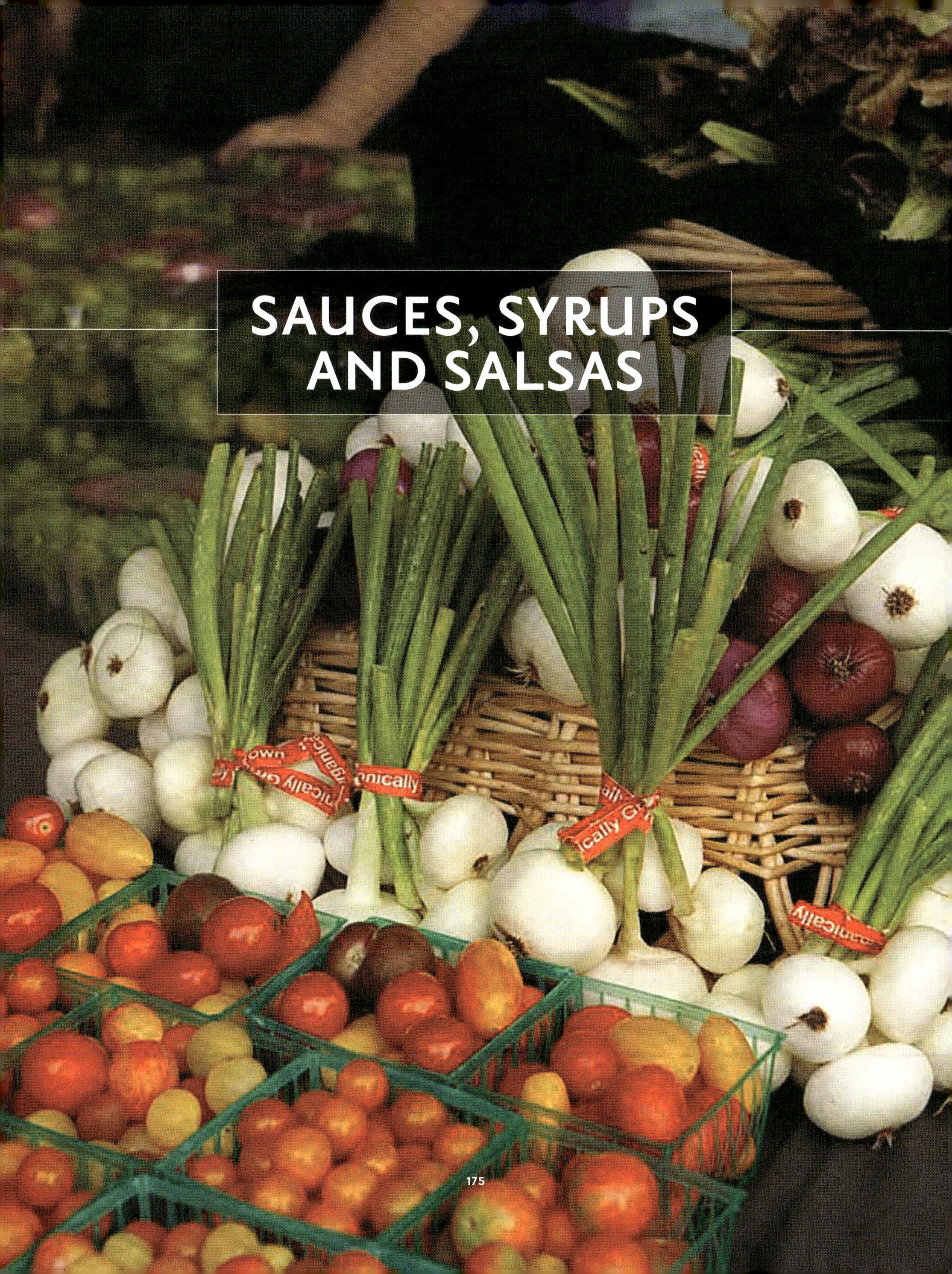

SAUCES, SYRUPS AND SALSAS

THREE WORLDLY SAUCES FOR GREEN BEANS AND ASPARAGUS (AND MORE)

Both asparagus and beans can have lengthy seasons in Wisconsin (that is, if you can call any growing season in the upper Midwest "lengthy"). When you find yourself running out of serving ideas, here are three savvy sauces from contributor Megan Bjella to keep them exciting. We'd bet any money that the sauces would go well with broccoli, carrots and cauliflower, too.

Sesame-Lime

3 tablespoons lime juice

1 roasted garlic clove or
1 teaspoon minced garlic

1 teaspoon grated lime zest

1 tablespoon toasted sesame oil

2 teaspoons honey

1/4 teaspoon or more
dried chile flakes

Garnish (optional): 1 tablespoon
toasted sesame seeds

Combine all ingredients except sesame seeds. Toss sauce with cooked vegetable of choice and garnish with sesame seeds.

Coconut-Tomato Curry

2 tablespoons neutral oil, coconut oil or ghee
(see page 140)

1 1/2 teaspoons mustard seeds

1 teaspoon cumin seeds, crushed in a
mortar and pestle

1/2 teaspoon ground turmeric

1/2 cup sliced onion

1 red serrano or jalapeño, seeded and thinly sliced

1 roasted garlic clove or 1 teaspoon minced garlic

1 1/2 cups chopped tomatoes

1/2 cup coconut milk

Salt

Heat oil and mustard seeds in a partially covered skillet over high heat. The seeds will start to pop as they fry; when this slows down, remove lid, reduce heat to medium and stir in cumin seeds, turmeric, onions and sliced chile. Cook until onions soften, 2 to 3 minutes. Stir in garlic, tomato and coconut milk; cook until tomatoes are saucy, 2 to 3 minutes longer. Taste and add salt as needed. Pour over cooked vegetable of choice.

Sweet-Spicy Tahini (or Peanut) Sauce

1 roasted garlic clove or 1 teaspoon minced garlic

1 tablespoon brown rice vinegar or rice vinegar

1 1/2 tablespoons tamari or soy sauce

3 tablespoons tahini or peanut butter

1 to 2 teaspoons sugar

1 teaspoon grated fresh ginger

Cayenne or dried chile flakes to taste

1 to 2 teaspoons toasted sesame oil

Whisk all ingredients in a bowl. Continue to whisk as you slowly add 1 to 2 tablespoons water, until sauce has thickened but is still pourable. Taste and adjust seasonings as desired. Use sauce to toss with—or drizzle it atop—cooked vegetables of your choice.

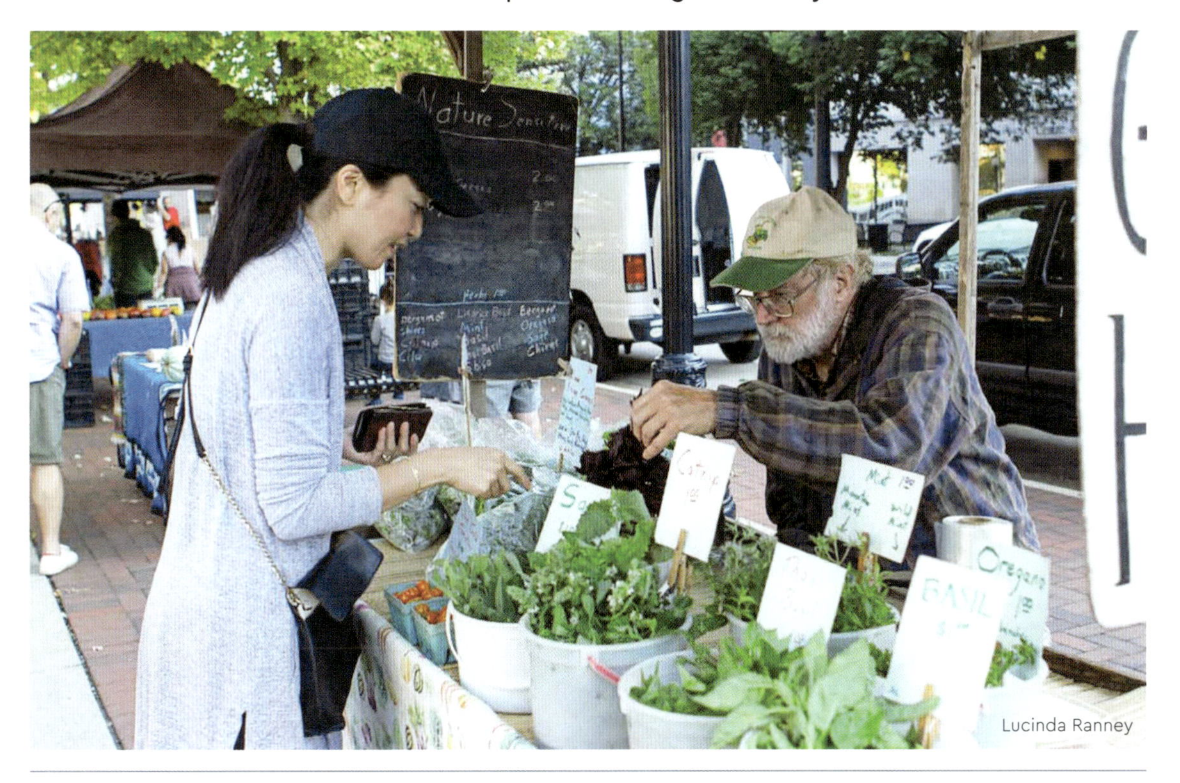

Lucinda Ranney

We'd bet any money that the sauces would go well with broccoli, carrots and cauliflower, too.

SWEET POTATO TOMATO SAUCE

MAKES ABOUT 4 CUPS

"I created this to satisfy a late-January pizza craving," says recipe contributor Clare Michaud. "I wanted to make a sauce with a warming depth of flavor, one that brought wintery ingredients to something with the bright summer flavor of tomatoes. This recipe helped me feel rejuvenated about cooking with root vegetables through Wisconsin's long winters."

A passionate cook whose cooking centers on seasonal ingredients, Clare is part of the market-day volunteer crew at the DCFM. "I love spending my Saturday mornings at the market, helping customers find a favorite vendor and getting to know the market vendors and their products."

Clare's sauce is easy to prepare, it keeps well in the fridge or freezer, and if you use vegetable stock, it's vegan. It can be tossed with pasta, used as a sauce for pizza or meatballs, or spread on sandwiches. Adaptations are welcome: Adjust the amount of stock based on how thick or thin you want the end results to be. Add sliced shallots to the roasting pan with the sweet potato and garlic. Use oregano instead of thyme, or sprinkle in some dried chile flakes. Clare makes it with the skins on the sweet potato, but of course you can peel it, if that's your preference.

1 large sweet potato (about 3/4 pound), peeled (or not) and diced (2 1/2 cups total)

4 whole medium garlic cloves

About 2 tablespoons olive oil, divided

1 teaspoon salt, plus more to taste

2 to 3 teaspoons fresh thyme leaves or 1 teaspoon dried thyme

1 pound ripe tomatoes, diced (about 2 1/2 cups) or 1 can (14.5 ounces) diced tomatoes

1 can (6 ounces) tomato paste

1/4 to 3/4 cup chicken or vegetable stock, divided

Black pepper

"I love spending my Saturday mornings at the market, helping customers find a favorite vendor and getting to know the market vendors and their products."

Heat oven to 400 degrees. Line a roasting or sheet pan with parchment, spread out the sweet potatoes and garlic cloves on it, then drizzle with 1 tablespoon of the olive oil and sprinkle with 1 teaspoon salt. Toss well, then roast mixture until vegetables are tender, stirring once or twice during the cooking, 20 to 30 minutes. Add the thyme (fresh or dried) when you stir the partially roasted vegetables.

In a blender or food processor, combine roasted sweet potato and garlic, tomatoes, tomato paste, 1/4 cup of the stock and remaining 1 tablespoon olive oil. Blend until smooth, adding additional stock to achieve desired texture. Add additional salt and pepper to taste. Use immediately, store in the refrigerator for up to a week, or freeze for future use.

Sarah Brooks

SMOKY EGGPLANT TOMATO SAUCE

3 TO 4 CUPS SAUCE

Market-goer Laura Schmidli makes an unusual tomato sauce that's thickened with fire-roasted eggplants, giving it a slightly sweet unctuousness. "My sister introduced me to the method of roasting eggplants skin-on over a flame, which makes [the flesh] very creamy and smoky." She uses an indoor method, one that's welcome when the weather is sweltering but the kitchen is blessedly air-conditioned.

Full-flavored on its own, Laura's sauce is also amenable to further seasoning to complement whatever use you're putting it into or on. Add fresh basil and garlic for an Italian pasta, for example, or za'atar to make a pita spread. Laura typically features it on homemade pizza (we're thinking oregano and chile flakes here).

If burning vegetable skin over an open flame isn't your thing (or your stovetop is electric), another method, albeit an outdoor one, is to slice the eggplant lengthwise into slabs, lay them skin side down on a hot gas grill and turn them occasionally during the cooking process. In this case it makes sense to roast the garlic and grill the tomatoes, too, using a grill pan, and again, turning them often.

The sauce will keep in the refrigerator for a week or in the freezer for several months.

2 to 3 large cloves garlic

5 cups cherry tomatoes

1/4 cup olive oil

2 to 3 large elongated Asian eggplants or 2 medium-small globe eggplants (1 1/2 to 2 pounds total), unpeeled

1/2 to 1 teaspoon smoked paprika

Salt and pepper

Preheat oven to 400 degrees. Wrap garlic in foil and bake until cloves are soft, 30 to 40 minutes. Also, coat tomatoes with the olive oil, place in a single layer in a baking pan and roast until golden brown, 20 to 30 minutes.

Meanwhile, pierce the eggplants a couple of times with a fork and then lay them on gas burner(s) over a medium-high flame or under a hot broiler. Char them evenly all over, rotating them with tongs periodically, about 12 to 20 minutes, depending on the size and type of eggplant. The skin should be blackened, the flesh should be soft, and the skin will begin to break.

Let eggplant(s) cool. Remove stem end and gently scrape away the blackened skin. (If there's a few specks of char left, no worries.)

When garlic, tomatoes and eggplant are done, place them in a blender or food processor with the paprika, plus salt and pepper to taste. Blend until smooth.

Sarah Brooks

"My sister introduced me to the method of roasting eggplants skin-on over a flame, which makes [the flesh] very creamy and smoky."

GARLIC SCAPE FETA PESTO

MAKES 2 GENEROUS CUPS

Americans first learned about pesto as a classic Italian cold sauce made with very specific ingredients: basil leaves, pine nuts, garlic, olive oil and Parmesan. But now we understand "pesto" as a broader term—one that welcomes other kinds of fresh herbs, nuts and cheeses into the mix. This version trades feta cheese for Parmesan and walnuts for pine nuts, and even the garlic has a change-up—instead of cloves, it features garlic scapes, those curlicued stems of early summer garlic that gardeners clip off so that the plant can send more energy to the bulbs.

"Garlic scape pesto is a seasonal favorite of ours," says contributor Jackie Gennett of Bushel & Peck's, a farm, shopping market and preservation kitchen all rolled into one business. Jackie purees the scapes with the first fresh basil of the season. "On our farm, we plant about two acres of garlic, so we have a lot of scapes. This is a fast, easy and delicious recipe that you can toss with any pasta. Or add a touch of orange cheese and some macaroni, and kids will love it, too." Be sure to try the pesto in the pasta salad recipe on page 80.

She recommends trimming off the slender tips of the scapes down to where the bud begins, since the tips can be stringy. To give the sauce some roastiness, we also suggest that you toast the walnuts before blending them with the other ingredients. Five to ten minutes in a 350-degree oven, tossing them once during the cooking time, will do the trick.

Garlic scapes can also be pickled, grilled or sauteed, notes Jackie.

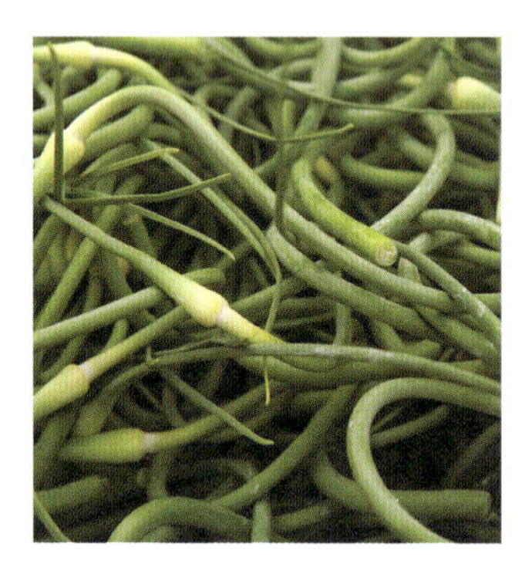

3/4 cup garlic scapes that have been chopped into 1-inch pieces

Packed 3/4 cup fresh basil leaves

1/2 cup olive oil

1/3 cup walnut pieces (toasted, if you have the time)

1/2 cup crumbled or chopped feta cheese

1/2 to 1 teaspoon lemon juice

1 teaspoon sea salt, or more to taste

Blend garlic scapes, basil and olive oil in a food processor until finely chopped. Scrape down the sides of the processor bowl with rubber spatula. Add remaining ingredients and blend until mixture is smooth. Add more salt to taste.

Lucinda Ranney

SALSA VERDE

MAKES 2 CUPS

Tricia Bross, of certified organic Luna Circle Farm, has been a member of the DCFM since 1990 (see photo on page 183). "As a young vegetable farmer, I wanted to grow everything. Tomatillos were something I had tasted in a salsa, but I had never grown them. It [turns out] that they grew very well. There I was with baskets and baskets of tomatillos, and nobody knew what they were, so they were not selling." Tricia put her business degree to good use and came up with the idea of a salsa kit containing tomatillos, onions and peppers from mild to extra hot. "Thirty-plus years later, my salsa verde baskets are still a hot seller. I like to think that I am responsible for introducing tomatillos to many folks in the Madison area."

The ingredient amounts called for in Tricia's salsa recipe are variable, so use the amount you like. For extra zip without additional heat, add a squeeze of lime at the end. For the freshest flavor, eat this soon after it's made; for a more blended, deeper taste, let it develop for one to three hours, in or out of the refrigerator.

Cooking notes: Here are some other options from Tricia to try:

Add 1 cup of sour cream to the salsa and stir. This will make the salsa thicker and creamier, and even more delicious because, well, that is the magic of sour cream.

Use the salsa as a sauce to cook chicken or pork chops. Brown the meat on both sides in a saucepan (I use a cast iron skillet). Then pour the salsa over the meat and continue to cook at low heat until the meat is cooked through. The salsa adds a great acidic tang that tenderizes and flavors the meat.

The salsa can be frozen for later use. Put it in a freezer safe container and pop in the freezer. When you thaw it for use, it will be a little watery. Just drain the excess water off. [Or drink it!]

1 hot pepper, or more to taste—choose the pepper that matches your heat preference

1 to 2 medium cloves garlic, peeled (optional)

1/2 bunch cilantro leaves, or to taste (optional)

1 pound (about 1 quart) tomatillos

1 medium red or sweet onion, finely chopped

Salt

"As a young vegetable farmer, I wanted to grow everything."

Remove the stem from the pepper and halve the pepper lengthwise. You can remove the seeds to reduce the heat of the pepper or leave them in for a spicier kick. Place pepper halves in a food processor with garlic and cilantro, if using. Process until the pepper is in small pieces.

Remove husks from tomatillos, add the tomatillos to the processor and pulse a few times to chop them, but don't overdo it. Scrape the mixture into a bowl and stir in the onions. Add salt to taste.

At this point you can sit in a comfy chair with a bag of tortilla chips and enjoy. Don't be surprised if it all disappears.

WE LIKE TO SAY "SALSA"

Everybody knows that salsa outsells ketchup in the USA. Born in ancient Mesoamerica as a combination of chiles, tomatoes, squash seeds and spices that was served with meat and fish, salsa became a signature sauce of Mexico that, when it crossed borders, evolved into an internationally beloved condiment with countless variations on a theme.

Today, there is a salsa for every taste, and we've got five of them to share with you.

Sweet n' Spicy Watermelon Salsa .... page 67

Sweet Corn and Bell Pepper Salsa .... page 74

Salsa Verde .... page 184

Ground Cherry and Tomato Salsa .... page 186

Oven Baked Salsa .... page 188

GROUND CHERRY AND TOMATO SALSA

MAKES ABOUT 2 CUPS

Sue and Todd Gronholz found their niche at the DCFM by specializing in the unusual. Once-unfamiliar items like heirloom tomatoes, wild black raspberries and anise hyssop lured customers to their stand from 1993 to 2005, when they retired from the market. One of their best sellers was ground cherries, the odd-looking, diminutive veggie that a writer for *Smithsonian Magazine* once described this way: "Tastes like a cherry tomato injected with mango and pineapple juice, and looks like an orange pearl encased in a miniature paper lantern."

Says Sue, "I can't tell you how many pints of ground cherries we sold over the years. Most people didn't know what they were, so we always had a container of samples; one bite and we usually gained a new customer. Our ground cherries were also popular with chefs, and some mornings they would buy us out before the market barely started."

This salsa is the result of her efforts to recreate a recipe shared by a chef-customer, who gave her ingredients but no measurements. Although regular basil is delicious in it, the pronounced licorice flavor of Thai or licorice basil, or anise hyssop is especially delicious. (See more about anise hyssop on page 78.) Serve it with tortilla chips or as a topping for grilled chicken.

2 tablespoons orange juice

1 tablespoon canola oil

1 tablespoon minced fresh basil, Thai basil, licorice basil or anise hyssop

1 teaspoon sugar

1 teaspoon orange zest

1/8 teaspoon salt (optional)

1 large tomato, seeded and diced (about 1 heaping cup)

1/2 sweet red pepper, diced (about 1/4 cup)

1/2 cup quartered ground cherries

2 tablespoons chopped sweet onion (such as Walla Walla)

1 or more tablespoons minced fresh jalapeño

Place orange juice, oil, basil or anise hyssop, sugar, orange zest and salt, if using, in a bowl and whisk to combine. Stir in diced tomato, red pepper, ground cherries, sweet onion and jalapeño. Refrigerate for at least 1 hour before serving, stirring occasionally.

"I can't tell you how many pints of ground cherries we sold over the years. Most people didn't know what they were, so we always had a container of samples; one bite and we usually gained a new customer.

OVEN BAKED SALSA

MAKES ABOUT 3 CUPS SALSA

Salsa-making time is summertime, but salsa-eating time is any season ... if you've cooked a batch for the freezer, that is. "I make multiple batches to last me for the next year," says contributor Diane Wiersema. "It truly uses so much produce from a farmers market, a CSA or one's own vegetable garden." She sometimes adds fresh corn scraped from the cob or leftover cooked corn to the recipe.

8 cups chopped tomatoes

2 cups chopped onions

1 cup chopped green peppers

8 seeded and chopped jalapeño peppers, or more or less to taste

1 to 2 tablespoons minced garlic

1/3 to 1/2 cup white or apple cider vinegar

1/4 cup honey

1 can (6 ounces) tomato paste

4 teaspoons salt

1 teaspoon black pepper

Heat oven to 325 degrees. Combine all ingredients in a large bowl and mix well. Add mixture to a deep 9-by-13-inch roasting pan and place the pan on a lined baking sheet to catch any drips. Bake until most of the moisture is gone and sauce has thickened to desired consistency, 2 to 3 hours.

Let salsa cool completely. Portion into glass jars and store in the refrigerator for up to a week, or in airtight plastic containers to freeze.

Char Thompson

Lucinda Ranney

PRESERVING THE ABUNDANCE: AMISH-INSPIRED APPLE CIDER SYRUP

The interdependence and camaraderie among chefs and growers can yield some very delicious results, as illustrated in this story from Odessa Piper, the James Beard Award-winning founder of Madison's celebrated L'Etoile restaurant. She relays a tale about apple cider syrup from L'Etoile's long and happy relationship with Weston's Antique Apples, a family orchard that specializes in old-time varieties.

> *It was late October after a hard freeze. Due to illness in the family, Ken Weston had not been able to get out into his orchard to finish the harvest. Many exquisite apples remained on the trees. The frozen apples were not fit for selling, but they could be turned into cider if we moved quickly enough. We made a special L'Etoile picnic for all who came to help and spent the rest of that very chilly day picking the trees clean. The Weston family got paid for their apples, and we made gallons and gallons of cider in their press.*
>
> *Then I had to figure out what to do with all of that cider! I took a cue from the thrifty Amish, who boil their cider into syrup for long storage. Over the next couple of days we reduced cider nonstop. The resulting syrup had a pleasing viscosity and a penetrating acidity and sweetness. It opened a profound new chapter in my culinary creativity [and] I found countless uses for it, from savory to sweet.*

A mix of distinctive flavors, colors and perfumes is what makes Weston's apple cider—and Odessa's syrup—so excellent. Below are her guidelines for making and using your own.

How Many Apples? How Much Cider?

Roughly speaking, 8 parts cider reduces to about 1 to 1 1/2 parts syrup. In cup-speak, that means that 8 cups (or a half-gallon of cider) reduces to 1 to 1 1/2 cups syrup. Early-season apples generally have more pectin, so those cider reductions will thicken to a pleasing viscosity at a ratio of 8 to 1 1/2. Later-season apples have more sugar but less pectin and need to be reduced further to a ratio of 8 to 1.

How to Reduce Cider into Syrup

Pour the cider into a heavy-bottomed pan that's big enough to leave extra room. Bring cider to a full boil over high heat, then reduce the heat to medium-high and maintain a medium boil. There's no need to stir. Monitor the reduction carefully and when it is reaching the desired volume ratio, reduce the heat to medium. Don't let the reduction caramelize because the apples' high-notes will be lost. Set a timer to remember to check every couple of minutes. During these final important minutes, test the reduction by spooning a drop onto a chilled dish (to get it to cool quickly). When a cooled drop has the viscosity of maple syrup, the cider reduction is ready. Cool it down, transfer it to a lidded container and refrigerate it. It will keep indefinitely.

Many Uses for Cider Syrup

- Combine equal parts cider syrup and miso paste and use the mixture to glaze a roast.
- Brush the syrup onto bacon, duck, pork or any fatty, rich roasting meat (except beef) during the last phase of cooking.
- Add a dash of cider syrup to Calvados, bacon, thyme and cream for a decadent sauce that's fantastic with chicken or seared scallops.
- Use cider syrup in place of some of the molasses for baked beans.
- Use the syrup to glaze roasting parsnips, turnips, rutabagas, beets, Brussels sprouts or any rich or naturally bitter vegetable.
- Mix with butter and bake in the well of a winter squash or an apple.
- Enrich the base of winter squash bisque, and drizzle some on the surface when you serve the soup.
- Use cider syrup as a base of barbecue sauce.
- Add a dollop of syrup when caramelizing sliced onions.
- Make a sweet-sour vinaigrette with cider syrup and cider vinegar. Use on rich salads that call for beets, nuts or blue cheese.
- Flavor whole-grain mustard with cider syrup and add chopped herbs for a custom condiment.
- Drizzle the cider syrup over maple-walnut ice cream.
- Pour cider syrup made from high-pectin, early-autumn vintage apples into small jelly jars. It makes a great gift that conveys a special time and place.

BLUEBERRY OPAL BASIL SYRUP

MAKES ABOUT 3 CUPS SYRUP

Fruit syrups may seem kind of fancy, but they're common in many cuisines. *Jallab*, for example, a popular beverage in the Middle East, is an iced tea that contains syrup made of grape molasses, dates and rose water. Apple cider syrup (see page 190) is part of Amish and New England's culinary heritage, traditionally used in baking cakes, pies and cookies.

This syrup, made from blueberries and an unusual basil varietal, was shared by former DCFM vendor Sue Gronholz. Sue serves it on panna cotta, pancakes, waffles and ice cream. When one of our recipe testers tried it out, she made julep cocktails, replacing the mint syrup in that classic Southern drink with this one. (See photo below.)

Cooking note: Opal basil has striking, dark-purple leaves, a heavenly fragrance and a milder flavor than sweet basil, with hints of cinnamon, anise, mint and clove. If you can't find opal basil at the farmers market, use a regular sweet basil, but about a half-cup less of it.

2 cups sugar

2 cups lightly packed fresh opal basil (leaves and stems)

1 cup blueberries (fresh or frozen, thawed and drained)

Place sugar and 2 cups water in a medium saucepan and bring to a boil over medium-high heat. Boil 8 to 10 minutes. Remove from heat, add basil and blueberries; cover and let steep until cool.

Strain syrup through several layers of cheesecloth, pressing solids with the back of a wooden spoon to remove as much liquid as possible. Discard solids; store syrup in the refrigerator. It will keep for 1 to 2 weeks.

Carole Blemker

Terese Allen

BAKED GOODS AND BREAKFAST DISHES

Recipe on page 198

Grant Johnston

BARA BRITH
(WELSH TEA BREAD)

MAKES 2 LOAVES, 8 TO 12 SLICES PER LOAF

Betty Rosengren, a DCFM fan for more than 30 years and counting, modified a recipe that her grandmother brought from Wales in 1917, when she first came to Wisconsin. In Welsh, *bara brith* refers to a traditional spiced tea bread that is speckled with fruit and served with afternoon tea.

"It was originally made with yeast, and there are many variations," Betty notes. "Dried candied citrus fruits, dried currants and fresh blackberries are some of the ingredients used." She likes to make the strong black tea that's called for with Twinings Prince of Wales, and her modifications include fresh red currants, sold by several vendors at the market. Tiny, jewel-like currants are available in July and can be frozen for later use. In the fall, you could also substitute chopped cranberries for the currants.

3/4 cup sultanas (golden raisins)

3/4 cup raisins

3 cups very strong hot black tea

3 cups flour

1 cup packed light brown sugar

3 3/4 teaspoons baking powder

1/2 teaspoon each ground cinnamon, allspice, nutmeg, cloves and coriander

1 1/4 teaspoons salt

1 1/4 cups whole milk

6 tablespoons unsalted butter, softened to room temperature

1 large egg, lightly beaten

1 cup fresh red currants

Tiny, jewel-like currants are available in July and can be frozen for later use.

Heat oven to 325 degrees. Oil or butter two 9-by-5-inch loaf pans. Place raisins and sultanas in a medium bowl, pour hot tea over them, stir, cover and let the mixture sit on the counter overnight.

Combine flour, brown sugar, baking powder, spices and salt in a large bowl. Stir in milk, softened butter and beaten egg until well mixed, about 1 minute. It should have a fairly smooth, medium-thick consistency.

Spoon off and reserve 2 to 3 tablespoons of the liquid from the soaked fruit, then drain off the rest of the liquid, if there is any. Stir the soaked, drained fruit into the batter. Gently stir in the red currants.

Divide batter into prepared loaf pans. Bake until a toothpick inserted near the center of the loaves comes out clean, 70 to 80 minutes. Remove bread from oven and drizzle the hot loaves with the reserved tea. Cool for 10 minutes, then remove bread from loaf pans and transfer them to a rack to cool completely.

The baked, cooled loaves should be wrapped and refrigerated, where they will keep three to four days. As the bread sits, its flavor deepens.

Betty Rosengren

Terese Allen

TO MARKET, TO MARKET TO BUY A FAT SCONE

If you love the cornucopia of fruits, vegetables, cheese and meats at the farmers market, get a load of the baked goods.

Bakeries lure hungry shoppers with apple fritters, lemon cream scones, spinach and cheese empanadas, raspberry muffins, filled croissants, cheddar biscuits—a veritable bonanza of breakfast pastries to nibble on as one strolls the stands.

But sweet rolls and savory hand pies are just the beginning of the bakeshop specialties available at the market. Artisan bakers offer naturally leavened, whole-grain loaves, dinner rolls and hamburger buns, plus a variety of granolas. Dessert artists create sparkling fruit tarts, fresh berry tortes and still-warm chocolate chunk cookies, packing them in cardboard boxes for take-home ease.

Over the past decade, gluten- and wheat-free treats have been added to the mix, as well as paleo, vegan, and celiac-safe choices. (We'll take a keto double fudge whoopie pie, please.) Most recently we've noticed an uptick in multicultural baked goods, such as flatbreads for easy pizza-making, chewy bagels encrusted with seeds and spices, olive-studded focaccia and dense loaves of dark ryebread.

The Dane County Farmers' Market mission is chiefly about promoting and selling Wisconsin-grown produce, but bakeries helped attract customers when the organization was young, and they've been an amenity at the market ever since. And through the years, bakers have added lots more regionally procured ingredients to their offerings—in such treats as apple cider donuts, cream cheese-frosted pumpkin bars and honey caramel cream puffs.

In order to set up on the inside ring of the Square, where market stands are located, a bakery must produce from-scratch goods and operate no more than one retail outlet, which in turn must be located where production occurs. At least one seller at the market must own a minimum of twenty percent of the business, and all of the owners must participate in the baking production. Because of such policies, the DCFM's bakeries, like its farms, maintain high standards and have a direct connection to customers.

The restrictions haven't slowed things a bit, either. From coconut macaroons and honey oat bread, to graham crackers and hazelnut shortbread bars, bakery is the "other" cornucopia of the market. Like the primary, vegetable-focused one, it overflows with variety, nourishment and simple pleasure.

TOMATO'S
BEANS
Fall at the
armers Market

YEAST-RAISED SWEET CORN BREAD

MAKES 3 LARGER OR 4 SMALLER LOAVES, DEPENDING ON SIZE OF PAN USED

Attention, bread-baking geeks: Here is a new, gotta-try recipe for you. Its creator is retired craft beer brewer Grant Johnston, an avid customer of the DCFM for many years. From Grant:

> *My original idea when formulating this bread, was to make full use of fantastic sweet corn during the summer high season. After baking this bread a few times, I began incorporating other ingredients that I thought would make a better loaf. Those included* masa harina, *fresh* pipicha *(a fantastic Mexican herb) and Mexican serrano or jalapeño chiles. I have also included cumin seeds that [are ground into a] powder in a coffee mill. Notice that the liquid in the recipe comes only from the sweet corn [and melted butter or oil]. This makes a very unique bread with a moist texture, and it keeps well."*

Grant's bread has a heady aroma, a fine chewiness and clear sweet corn flavor. It's particularly good with BLTs (so when you're at the market buying sweet corn, pick up some B, L and T, also.) The loaves freeze well, plus they are very gift-worthy, so if your household is small don't let the three- to four-loaf yield concern you.

This is a project recipe; it involves time, effort and at least some familiarity with bread baking. But try it out even if you're not an experienced baker—Grant has supplied expert guidance to lead you down the yeasty road, covering everything from flour brands and pan size to the benefits of a slow rise and why weights are better than measurements. Before you begin, be sure to read through "Notes for the Baker" on page 201 and the recipe itself.

Cooking note: For the sweet corn puree, if you're using the measurement instead of the weight, be sure to measure *after* the corn has been pureed.

Grant Johnston

1/2 teaspoon active dry yeast, 1/3 teaspoon instant yeast or 3 grams fresh yeast (do not use fast-rise yeast)

1,200 grams (about 6 cups) uber-fresh, well-pureed sweet corn (from about 8 ears)

50 grams (about 8 tablespoons) very finely chopped fresh pipicha or tarragon

50 grams (4 tablespoons) melted unsalted butter or
50 grams (about 3 1/2 tablespoons) corn oil

10 grams cumin seeds (about 1 tablespoon), ground to a fine powder in a coffee mill

15 grams (about 1 tablespoon) finely minced fresh serrano or jalapeño (optional)

21 grams (about 4 teaspoons) fine sea salt

800 grams (about 7 cups) bread flour

50 grams (about 1/2 cup) whole wheat flour

50 grams (about 1/2 cup) whole rye flour

100 grams (about 3/4 cup) Maseca™ instant yellow or blue corn masa flour

Corn oil for oiling the dough-rising container and the dough

Softened butter for buttering the bread pans

Heat 1/4 cup of water to 80 degrees (use an instant-read thermometer to test water temperature), add yeast and mix well to dissolve. Set aside for 10 minutes.

Meanwhile, place the pureed corn, pipicha or tarragon, melted butter or corn oil, ground cumin, minced chiles and salt into a very large mixing bowl. Stir until salt is dissolved and everything is mixed well, 1 to 2 minutes. After yeast mixture has stood for 10 minutes, add it to the corn mixture and stir well to combine.

Whisk all four types of flour in a separate bowl until they are thoroughly combined. Remove 3/4 cup of this flour mixture to a bowl and set it aside. Add remaining flour mixture to the liquids in the large bowl and use a large, sturdy wooden spoon to mix everything very well, leaving no dry flour spots. The dough should be a bit sticky at this point. Let dough rest in the bowl 30 minutes so that the flour has time to absorb all the liquid. (Do not skip this step!)

Lightly dust the work space with a little of the reserved flour mixture. Turn the dough out onto it and knead it for 5 minutes, adding some or all of remaining reserved flour mixture only as needed, until the dough is smooth and holds together as a well-formed ball.

Use corn oil to generously oil a bowl or container that is large enough to hold twice the amount of dough. Add dough to bowl and oil the surface of the dough, too. Loosely cover the bowl with plastic wrap and place it in a cool room (between 65 and 70 degrees). Let the dough slow-rise until it has doubled in size (a full 100% increase in volume), 8 to 10 hours, or overnight.

Transfer the dough to a floured work surface and gently deflate it to get rid of most of the carbon dioxide that has formed in the dough. Cover it loosely with plastic wrap and let it rest 15 minutes.

Knead dough again briefly, about 1 minute, to redistribute the yeast and to maintain even dough temperature. Reform it into a ball and place dough back into the oiled bowl. Cover bowl as before and let it rise again in a cool location until doubled in size. This second rise will be much shorter than the first, 3 to 4 hours or so. But remember: The amount of rising time will vary, depending on temperature. Watch the dough, not the clock.

Turn dough out onto a lightly floured surface, gently deflate it and let it rest 15 minutes. Meanwhile, generously butter three standard American bread pans that are 8 1/2-by-4 1/2-by-2 1/2 inches in size, or four British farmhouse tins that measure 6-by-3 1/2-by-2.8 inches in size.

If you're using the larger American pans, use a sharp knife to cut the dough into three pieces of approximately equal weight—about 800 grams (or 1 3/4 pounds) each. If you're using the smaller British tins, cut dough into four pieces that are about 600 grams (or about 1 1/3 pounds) each. Knead each piece briefly—about 1 minute—and form each one into an oblong loaf to fit the pan size.

Place formed loaves seam side down in the pans. Cover them, this time with a very lightweight tea towel (so that it won't stick to the loaves). Let dough rise again in a cool location until it has risen about 3/4-inch above the top of the pans, about 1 1/2 to 2 hours, depending on room temperature. The amount that the dough rises is what matters at this point. The photo on the left shows when the dough is ready for the oven.

Place oven rack on the lower-third shelf of the oven. Heat oven to 375 degrees. Bake for 10 minutes, then loosely cover the loaves with foil to prevent excess browning. Continue baking until an instant-read thermometer inserted into the bread reads 205 degrees, about 40 minutes.

Remove foil and continue to bake bread until it's medium-dark brown; this will take only a few minutes. Watch carefully so bread doesn't brown too much—this happens quickly.

Remove bread from oven and turn the loaves out of the pans onto cooling racks. (If the bread needs coaxing here, use a small, flexible, metal spatula or thin-bladed knife to loosen it around the sides.) Be patient! This is a very moist bread, so it's imperative that it cools completely before slicing—it is very worth the wait.

Photos Grant Johnston

NOTES FOR THE BAKER: YEAST-RAISED SWEET CORN BREAD

- Pipicha is an aromatic Mexican herb with a flavor that resembles cilantro with a citrus or lemon taste. If you can't get fresh pipicha from a farmers market or Mexican grocery store, use fresh tarragon instead. Tarragon is very different than pipicha but also delicious in the bread.

- Choose the freshest possible sweet corn for this bread so that one of the finest flavors of summer will be front and center. If you freeze corn at its absolute peak and then thaw it for this recipe, the bread can be baked and enjoyed year-round.

- For topnotch results, use Meadowlark Organics or King Arthur flours, and Maseca™ instant masa corn flour here. If you're not able to get those brands, look for the best and freshest flours available. Remember, different flour types and brands absorb different amounts of liquid, so use the volume measurements given here as a rough guide. Using an accurate digital scale to measure by weight is far preferable.

- This recipe uses far less yeast than do most recipes. This is by design, and it will take longer for the dough to rise. This is a good thing because slower, cooler rises make for much better-tasting and longer-keeping loaves. Rising times for the dough given are estimates only. Watch the dough and not the clock.

- Another variable that affects rising time is room temperature—a warmer room causes a faster rise. If your dough is rising too quickly, whether because of temperature or the yeast, you can move it to the refrigerator to slow things down.

Grant Johnston

- A clear, plastic, box-shaped container is ideal for dough-rising, as it is easy to see when the dough has doubled in volume. They're available at restaurant supply stores and online.

- Pan size is important. For this recipe, use standard American bread pans that are 8 1/2-by-4 1/2-by-2 1/2 inches in size, to yield 3 larger loaves. Or—even better, use British pans that are called farmhouse tins and are 6-by-3 1/2-by-2.8 inches in size. They have high sides and make a much better formed loaf, particularly with this dough. Whatever you do, don't use loaf pans that are 9-by-5-inches; they're too big and are the wrong shape for this bread.

- A kitchen item that's especially helpful in this recipe is an accurate, digital, instant-read thermometer to test water temperature for dissolving the yeast and to test the bread for doneness.

CHEESY WHITE CORN GRITS

4 SERVINGS

Grits are a beloved breakfast standard in U.S. South, but not so much in the upper half of the country. That might change, however, if more Northerners were to try a steaming bowlful made with Meadowlark Organics' locally grown, freshly milled white corn (see page 201).

Meadowlark co-owner Halee Wepking told us, "Dent corns have historically been the variety used to make grits, and our white corn grits are milled from a variety of open-pollinated white dent corn on our farm in Ridgeway, Wisconsin. Because this corn is open-pollinated, we are able to save seeds each year, versus a hybrid corn which will not produce another crop [that is] true to type. This means that each year we walk the fields to pick the best ears we can find to save for planting the following year."

For corn destined to become grits, she says, "We mill it on a designated mill that produces a coarse cornmeal, with the powdery corn flour sifted off." The resulting dish has a roasty, deeply corn-forward flavor and a rich-feeling texture. And once you add a good Wisconsin-made sharp cheddar to the dish, well, you'll have yourself a new breakfast standard. "While cheddar is very traditional, feel free to experiment with your favorite melty cheese," says Halee.

Here are her serving ideas: "Topped with a fried egg and doused in your favorite hot sauce; with shrimp for a classic Southern variation; with stewed or sauteed greens; or topped with shredded pork and salsa verde." For the latter, try it with the Carnitas and Salsa Verde recipes on page 120 and 184, respectively.

Breakfast standard? Make that a breakfast, lunch and dinner standard.

Cooking note: Using water as the liquid here really lets the corn flavor come through. For a different savor, substitute 4 cups chicken stock, or 2 cups water and 2 cups milk, for the water.

1 cup Meadowlark Organics White Corn Grits

1 teaspoon minced garlic, or more as you like

4 tablespoons butter

2 tablespoons heavy cream

4 ounces (about 1 cup) grated sharp cheddar

Garnish (optional): Chopped chives

Bring 4 cups water to a boil over medium-high heat in a saucepan. Add the garlic and slowly whisk in the grits (using a whisk helps prevent lumps as the mixture comes to a boil and begins to thicken).

Turn heat to low, switch from a whisk to a wooden spoon and stir grits every 10 minutes or so as they cook, for about 40 minutes, adding a splash of water now and then as necessary to keep grits from thickening too much. It may take as much as 2 cups more liquid to achieve a smooth, pourable consistency. (If you find the grits are sticking to the bottom of the pan, remove pan from the burner and let it sit a few minutes. Scrape the bottom of the pan with your wooden spoon, stirring to incorporate, and return it to the heat.)

Stir in the butter, heavy cream and cheddar. Once the cheese has melted, add salt and pepper to taste, and serve it up, garnished with a sprinkling of chives, if you like.

Rachel Figuerod

GRANDMA'S DUTCH PANCAKES

12 EIGHT-INCH PANCAKES

Pancakes come in styles, from oniony potato latkes and spongy *injera*, to Russian *blini* and cornmeal johnnycakes. The pancakes here hail from the Netherlands; they're eggy and thin, but not as thin as a French crepe, and are served rolled or flat with sweet or savory additions.

Don't confuse them with Dutch baby pancakes, which are oven-baked and puffy, and not Dutch at all, but rather German. The term "Dutch baby" came from an American restaurant wherein the use of "Dutch" was a mispronunciation of *Deutsch*, the German word for "German." "Baby" referred to the small size of the pancakes.

This recipe comes from Madison-based journalist Christina Lorey. "In my family, we call these 'Grandma Lorey's pancakes,' after my dad's mom, who would spend hours making them for her husband, kids and their families. Grandma Lorey died when I was just one, so I really got to know her through her recipes—mainly these Dutch pancakes."

When it comes to the toppings, "Creativity is welcome," says Christina. Let the market inspire you here or choose from her suggestions below. "If you're cooking for a group, it's great to have multiple skillets going. Or start early, prepare a bunch of pancakes, and keep them warm on a tray in your oven at the lowest temperature possible."

8 eggs

1 cup milk

2 cups flour

1/4 cup sugar

1 teaspoon vanilla extract

Butter for cooking pancakes

For serving (choose from the following):

Softened butter

Wisconsin-made maple syrup

Any farmers market fruit or jam/jelly

Fresh berries or other market fruits

Powdered sugar

Fried eggs (over-easy or as you like)

Cooked bacon strips

Warm a well-loved 8-inch cast iron skillet over medium heat. Meanwhile, make the pancake batter: Mix eggs, milk, flour, sugar and vanilla together. Beat well (use a hand mixer if you like) until you see air bubbles in the batter. The batter will be thin.

Add a dab of butter to the skillet and swirl pan to butter the entire bottom as it heats. Then pour 1/4 cup of batter into the skillet, again swirling the pan to make sure batter fills the entire area inside the skillet. Let it cook on the first side until batter no longer looks wet, less than 2 minutes (adjust the heat as needed). Then use a thin spatula to flip the pancake over. After 30 more seconds, remove pancake from skillet. Repeat with remaining batter, adding a bit more butter each time. Serve immediately, with any of the listed suggestions.

Terese Allen

SHAKSHUKA

4 TO 6 SERVINGS

From contributor Efrat Livny: "Said to have originated in North Africa, the *shakshuka* is a simple dish made of gently poached eggs in a thick tomato and bell pepper sauce. It's primarily a breakfast dish but can be served and eaten any time of day. It is sometimes prepared in a large skillet and dished out, but often it is made in a personal-sized pan and brought to the table right off the burner, still bubbling. The latter allows [each diner] to select the level of heat in the dish and add preferred condiments, such as parsley or feta."

Efrat has been a faithful DCFM shopper since she arrived in Madison in 1983. Forty years later, she's still "getting up early and rushing downtown with a sense of excitement, rain or shine. I always say that I cook food grown by people I hug, which is the true secret of a long and happy life."

Serve Efrat's shakshuka with fresh pita or any hearty bread for sopping up the sauce and egg yolk. "A tomato and cucumber salad with a light olive oil and lemon dressing or tahini sauce and some chopped mint is a traditional and refreshing side to serve along a hot shakshuka."

For more about shakshuka, see page 208.

2 medium onions

6 medium cloves garlic

2 tablespoons olive oil

2 red bell peppers

6 large ripe tomatoes

1 to 2 tablespoons ground cumin, divided

1 teaspoon sweet paprika

1 to 2 teaspoons salt, divided

1/2 teaspoon freshly ground black pepper

Pinch of cayenne pepper, Aleppo pepper or harissa powder (optional)

1 to 2 tablespoons tomato paste, if needed to thicken the sauce

8 large eggs

About 1 cup crumbled feta, preferably Bulgarian or another creamy variety (optional)

Garnish (optional): 2 tablespoons chopped cilantro or parsley

Cut the onions into 1/4-inch pieces (about 1 1/2 cups total). Thinly slice the garlic (about 1/4 cup). Heat the olive oil in a large, deep skillet that has a lid, then add the onions and saute them over medium heat until golden. Stir in the garlic and cook until everything is lightly browned, about 10 minutes.

While the onions are cooking, cut the peppers into 1/2-inch pieces (about 1 cup) and add them to the skillet. Stir the mixture occasionally as the peppers soften. Cut the tomatoes into 1-inch chunks (about 4 cups) and add to the skillet. Cook the mixture until the liquid that releases from the tomatoes partially evaporates, 10 to 12 minutes.

Stir in 1 tablespoon of the cumin, the paprika, 1 teaspoon of the salt and the black pepper. Reduce heat and continue cooking, stirring occasionally, until mixture is thick, another 25 to 35 minutes. (If mixture becomes too thick, add a little water; if it remains too thin, stir in 1 to 2 tablespoons tomato paste.)

Taste the sauce and add more cumin and salt, if desired, and add the cayenne or other pepper seasoning, if using. At this point, the sauce can be cooled and refrigerated for later use.

To finish the shakshuka: Reheat the sauce in the large skillet. When it's hot, make 8 small, evenly distributed indentations in the mixture. Break a single egg carefully into each of the indentations. Cover skillet and cook until the egg whites are solid but the yolks are still runny, 5 to 10 minutes. Crumbled feta can be added on top of the sauce in the areas between the eggs and allowed to melt during the last couple of minutes of cooking. Or it can be offered as a side condiment for each person to add to their portion as desired. Sprinkle shakshuka with cilantro or parsley, if desired, and serve.

Terese Allen

A DISH MADE FOR MEMORIES

I grew up and learned to cook in Israel. While my ancestry is European, my parents loved local street food that was influenced by our neighboring Arab countries—Egypt, Lebanon and Jordan—and food traditions brought by Sephardic Jews from North Africa, Turkey, Yemen, Iraq and Spain. Amongst the most popular foods, such as falafel and hummus, was the shakshuka. I often was taken on weekly excursions to the large food market in Jerusalem to get a weekly supply of vegetables, fruit, meat, fish, pickles, cheese and various condiments. When purchases were completed, laden with full baskets and with the calls of the merchants still ringing around us, we would go to a small restaurant and enjoy these staples, freshly made and bursting with flavor.

My favorite shakshuka memory is from a side street in Istanbul, Turkey, where I chanced upon makeshift low tables and long benches set up on the sidewalk by the door of a tiny restaurant. I joined the local crowd of late-morning diners who were being served individual portions of shakshuka in handle-less frying pans that were covered with soot from a woodburning oven and set near piles of large fresh pita breads. I can still see, smell and taste that tongue-burning amazingness!

Classic shakshuka has taken on many variations and elaborations. Sometimes ingredients such as eggplant, zucchini and/or spinach are added to the [traditional tomato-pepper sauce]. However, some shakshukas are further transformed by eliminating the tomatoes and red peppers and focusing on green ingredients like zucchini, chard, asparagus, kale, brussels sprouts, broccoli, spinach and various herbs. Mushrooms are also sometimes added.

I make shakshuka often for my family and friends, old and new. I think that many of them consider it my signature dish. It provides me a way to bring memories, connections and experiences from my beloved and first homeland into my present home and community.

—Efrat Livny, Madison

Efrat Livny

Terese Allen

DINOSAUR EGG OMELET

1 SERVING

In case you're wondering, you won't find dinosaur eggs at the farmers market, and you won't find them in this recipe, either. "Dinosaur" herein refers to the size and look of the omelet as it cooks, not to the type of eggs used—which, ideally, are pasture-raised chicken eggs, the gold-yolked, nutrition-packed kind you *can* get at the market.

Contributor Daniel Tortorice, a DCFM customer since its creation in 1972, is a big believer in using breakfast as a way to charge up his day. He began adding egg whites to his diet years ago to improve his blood pressure and cholesterol levels. If cholesterol is not a concern for you, consider this a go-to recipe when you find yourself with leftover egg whites after making a recipe that calls for yolks only.

"The problem with omelets," says Daniel, "is that the insides are dry, so we add rich sauces and other fatty ingredients to get some moisture and flavor. This method has its own sauce: the runny egg yolk." As for the filling, there's a near-endless number of multicultural possibilities. Daniel offers some ideas on the following page, and we tacked on a few more. Better get cracking.

Cooking note: Daniel uses unbeaten egg whites for the base of this omelet, which visually are easy to assess for doneness and yield an elegantly smooth, pale-yellow-on-the-outside omelet (see photos below). We also tried it out with whipped whites, which took longer to cook and was harder to assess for doneness, but made a thicker, airy, almost "frittata-like" omelet. Give both styles a whirl, if you like, and see which one you prefer.

Photos Ruth Bronston

3/4 to 1 teaspoon extra-virgin olive oil

2 large unbeaten egg whites

1 large egg

Salt and pepper to taste

Market ingredients of your choice, such as:

- Parsley, dill, tarragon and basil, and finely crumbled feta cheese
- Crumbled bacon and watercress leaves
- Sauteed chanterelles, oyster mushrooms or morels
- Chipotles and queso fresco
- Baby spinach and goat cheese
- Salsa and cooked, crumbled chorizo
- Kimchi and cilantro
- Fine-chopped, lightly cooked asparagus and ramps
- Pea or sunflower sprouts and slivered green onions
- Diced tomatoes and blue cheese
- Sauteed slivered Brussel sprouts, fresh thyme and Parmesan
- Any kind of pesto

Place an 8-inch, sloped, nonstick skillet over a low flame. Coat the bottom of the pan with olive oil. Gently warm the oil, then add the egg whites. Let them begin to set for 10 to 15 seconds, then break the whole egg into a small cup and slide the whole egg onto the egg whites. Use a rubber spatula to gently center the yolk in the middle of the pan. Tilt the pan a little this way and that, allowing the whites to set a bit around the yolk and hold it in place. Sprinkle with salt and pepper.

Cover the pan to trap steam inside, preferably with a glass lid so you can watch the eggs as they cook. Cook mixture until the eggs are done but the yolk is still runny, about 4 to 5 minutes. It will look like one giant egg. Now add your filling choices to the "dinosaur egg," being careful not to break the yolk (yet). Slide the egg out onto a plate, folding it over so it becomes an omelet. Serve immediately.

As for the filling, there's a near-endless number of multicultural possibilities.

Lucinda Ranney

DESSERTS

ROSH HASHANAH HONEY APPLE BUNDT CAKE (AND A GLUTEN-FREE VERSION)

16 OR MORE SERVINGS

This handsome, meaningful cake is a family favorite from DCFM shopper Betsy Abramson. Betsy tells us, "Honey cake is the traditional dessert for Rosh Hashanah, the Jewish New Year, which is in fall every year. [Another tradition] is to dip apples into honey before you start the meal—honey is to represent our hopes for a sweet new year, and the apple represents the earth's blessings and that the year goes 'round just like the round apple."

Betsy likes using honey *and* apples in her honey cake, in part because both primary ingredients are available at the DCFM. "To me, DCFM honey plus apples plus fall equals Rosh Hashanah."

The original recipe called for Granny Smith apples, which are perfect if you're in a pinch and it's the only firm-fleshed, tart eating/baking apple you can find at the grocery store. At the market, however, you'll find many locally grown varieties that have similar qualities, including Northern Spy, Newtown Pippen, McIntosh and Cortland. If you're wondering how to tell if what you're buying is the type you want, just ask the grower. (This is one of the reasons we love the farmers market, right?)

Cooking notes:
To make a gluten-free version of this cake, substitute Bob's Red Mill Gluten Free 1-to-1 Baking Flour for the wheat flour and use 2/3 cup honey instead of 3/4 cup.

Sprinkle the cake with powdered sugar just before—or not too long before—serving the dessert. (Otherwise, says Betsy, "the cake is so moist that the powdered sugar will all sink right into it.") Or you can drizzle vanilla-flavored powdered sugar icing over it, too.

Oil for greasing pan

3 large eggs

3/4 cup honey

1/2 cup white sugar

1/2 cup light brown sugar

1 1/4 cup vegetable or canola oil

1 1/2 teaspoon vanilla extract

3 cups flour

1 teaspoon baking powder

1 teaspoon baking soda

3/4 teaspoon salt

1 1/2 teaspoons cinnamon

1/4 teaspoon allspice

Dash of ground cloves

5 medium firm-fleshed tart apples (such as Northern Spy, Newtown Pippen, Macintosh, Cortland or Granny Smith)

Garnish (optional): Powdered sugar

Heat oven to 325 degrees. Oil a Bundt pan or other fluted cake pan (one that is 9 1/2 to 10 inches across at the top), taking care to evenly coat the nooks and crannies of the entire interior surface.

Crack eggs into a large mixing bowl and whisk until frothy. Whisk in honey, sugars, oil and vanilla until the mixture is well-combined and thickened. (Alternatively, you can use electric beaters.) Set aside.

Sift flour, baking powder, baking soda, salt and spices into a second, smaller bowl. Set aside.

Peel, core and shred the apples on the large holes of a box grater (or use a food processor). There should be 4 1/2 to 5 loosely packed cups. If the shredded apples are very juicy, place them in a clean towel and squeeze to remove some moisture.

Gently stir the flour mixture into the liquid mixture, taking care not to over blend it. (A few streaks of flour are fine.) Fold in the shredded apples just until they're well-distributed—don't go any further. Scrape the batter into the Bundt pan—it should not be more than three-quarters full. Use a rubber spatula to gently ease the batter toward the outsides of the pan, and very gently push slightly up the walls, to rid it of any air pockets. Now smooth the top so it's flat and level.

Bake the cake 50 minutes and check it for doneness—the edges should be darkened and pulling fully away from the sides of the pan, and the cake browned all across the surface. It may take another 20 to 30 minutes for this, depending on whether you're using a light- or dark-colored pan (dark-colored bakes faster), and how accurate your oven is (home ovens can often be 5 to 25 degrees off). You can also use a slender skewer to test for doneness, inserting it into the deepest part of the cake and withdrawing to check that it comes out clean.

Ruth Bronston

When it's done, remove cake from the oven and let it cool on a rack for at least ten minutes, then invert it onto a flat plate or platter, tapping the Bundt pan gently to release it (and praying, too). If the cake sticks, use a plastic knife to gently and carefully loosen the cake around both the center tube and sides. Let it cool completely before serving. Sift powdered sugar over the cake just before serving.

LANCASHIRE COURTING CAKE

ABOUT 16 SERVINGS

Joan and the late John Oosterwyk began selling farm-fresh fruits and vegetables at the DCFM in 1984, but over the years it was Land of O's jams, jellies and bottled goodies that became their trademark. No surprise there, what with Joan's carrot marmalade, lemon ginger pear jelly, cranberry horseradish relish and such, all lined up at their booth like glittering baubles in a jewelry display case. Today, the second generation has taken over on the farm and at the market. The Land of O's legacy lives on.

Market regular Ally Shepherd uses Land of O's Raspberry Strawberry Jam in this lovely tea cake, a treat she remembers from her childhood in the U.K. She writes, "Courting cake was traditionally made in a working-class part of England called Lancashire. Women wanting to impress their beloveds with their skills in the kitchen would make it to show what a good wife they would make (hence the name)."

The dessert's texture is "somewhere between a cake and a biscuit—or what you call cookies." (Americans would probably describe it as a bar.) It's okay to use "very ripe strawberries instead of jam in the cake, or use another jam of your choosing," says Ally. And of course, "Serve with a cup of English tea." We think the Oosterwyks would approve.

Cooking note: During Ally's grandmother's time, margarine was often used in baking to save money. It makes a fine pastry here, but butter gives a sweeter and, well, more buttery taste.

Ally Shepherd

Terese Allen

1/2 pound (2 sticks) butter or margarine, cool but not cold, cut into small pieces, plus a little extra to grease pan

3 1/3 cups (1 pound) flour

1 cup sugar

1/2 teaspoon baking soda

2 eggs, beaten

About 3 cups jam (raspberry, strawberry, a combination of those two, or your favorite jam)

Milk

Place butter or margarine, flour, sugar and baking soda in a large bowl and rub the ingredients together with your fingertips until mixture resembles breadcrumbs. Mix in the eggs until everything is well combined and a dough has formed.

Heat oven to 375 degrees. Use a little butter or margarine to grease an 8-by-11-inch or 9-by-9-inch baking pan with sides. Lay a 6- or 7-inch-wide piece of parchment paper into the pan, one long enough so that there's extra paper hanging over two opposite sides of the pan. (This will be used later to lift the cake from the pan). Grease the parchment paper, too.

Terese Allen

Transfer half the dough to a large piece of parchment paper on a work surface. Place another piece of parchment over the dough and then roll out dough to the size that will fit into the bottom of the prepared pan. Use the parchment to lift dough and transfer it to the pan, then remove dough from parchment and fit into place. Spread jam on the dough to a thickness of about 1/3 inch. Roll out the other half of the dough between the parchment sheets to fit the pan and then place it over the jam.

Brush the top with milk. Bake 15 minutes, then turn it around in the oven and continue baking until it's lightly browned, another 12 to 15 minutes.

Remove from oven and let it cool in the pan about 10 minutes. Run a sharp knife around the edges to loosen the cake, then use the parchment paper edges in the pan to lift cake from the pan and onto a cooling rack. Let it cool at least another 30 to 40 minutes. To serve, cut cake into finger-like rectangular pieces.

BAVARIAN FRUIT TORTE

8 SERVINGS

Tortes are prevalent in Germany, Hungary and Austria, but to be honest, this is not really one of them. European tortes are cakes layered with buttercream, jams, mousse and other fillings, often tall affairs with sweet icing on top and/or on the sides. Tarts are flatter, more like open-faced pies. They typically have a pastry or shortbread crust and are filled with fruit, custard or the like—which sounds more like what we have going here. Yet search the web for "Bavarian Torte" and this dessert is pretty much what pops up. So torte it is.

The recipe is adapted from a printed handout distributed by Ten Eyck Orchard, the venerable, sixth-generation operation located near Brodhead, Wisconsin. Besides 50 varieties of hand-picked, sustainably raised apples, they also offer pears, pumpkins, squash, cider, apple butter, honey and made-from-scratch baked goods.

While the recipe on the Ten Eyck handout called for apples, we tested the torte with pears and switched the vanilla extract to almond (just for the fun of it). But you can switch those right back if you'd like—or substitute other fall fruits, like plums or cranberries, for the pomes. With a foolproof mix-and-press crust and a fruit-topped cream cheese filling, it's a cook-friendly keeper of a recipe—whether you call it a torte or a tart.

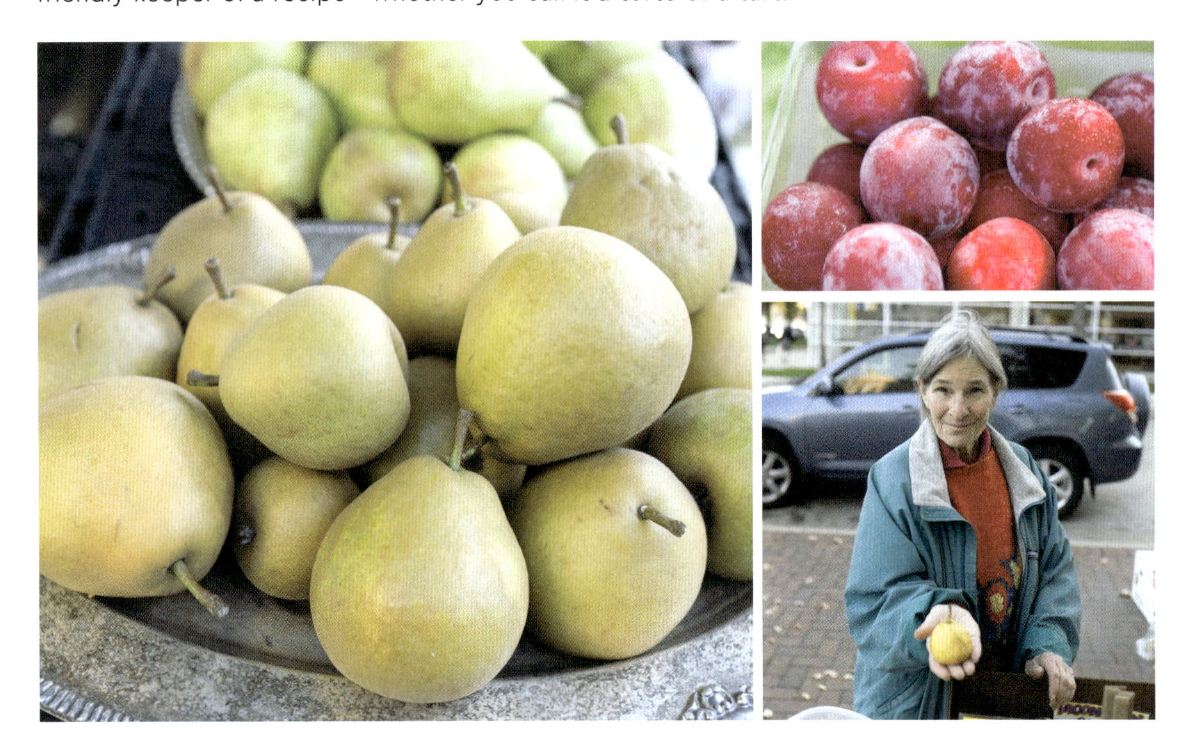

Crust:

1/2 cup (1 stick) butter, softened to room temperature

1/4 cup sugar

1/4 teaspoon almond extract

1 cup flour

Topping:

6 to 8 medium ripe pears, apples or other fall fruits (1 1/2 to 2 pounds)

1/4 cup sugar

1/2 teaspoon ground cinnamon

1/4 cup sliced almonds

Filling:

1 package (8 ounces) cream cheese, softened to room temperature

1 egg

1/4 cup sugar

1/2 teaspoon almond extract

Heat the oven to 425 degrees.

Make the dough for the crust: Use a wooden spoon to cream the butter, sugar and almond extract in a bowl until smooth, 1 to 2 minutes. Stir in the flour until a soft, smooth dough forms. Transfer the dough to a 9- to 9 1/2-inch springform pan and pat it evenly into the bottom of the pan. Set aside.

Make the filling: Wipe out the bowl you just used and add the cream cheese, egg, sugar and almond extract to it. Using a whisk or electric beaters, beat the mixture until smooth, 1 to 2 minutes. Set aside.

Make the topping: Peel, core and thickly slice the pears or other fruit. Toss them in a bowl with the sugar, cinnamon and almonds.

Assemble and bake the torte: Scrape the filling into the prepared crust and smooth it out with a spatula. Gently spoon the topping over the filling, covering the surface evenly. Bake the torte for 10 minutes, then reduce the heat to 400 and continue baking until the filling is set and the fruit is a little bubbly, about 25 to 35 minutes.

Cool the torte completely before serving it. It's good at room temperature or chilled. (If there are any leftover slices, they're terrific heated in a 325-degree toaster oven for about 5 minutes.)

With a foolproof mix-and-press crust and a fruit-topped cream cheese filling, it's a cook-friendly keeper of a recipe.

STRAWBERRY RHUBARB DESSERT BARS

12 TO 16 SERVINGS

Strawberry-rhubarb pie is one of the early summer darlings of the Wisconsin table, but this dessert bar gives you all the pleasure of pie without the crust-making anxiety. The pastry bottom is buttery, the topping stays gooey and the edges caramelize deliciously. Who needs pie crust?

What's more, the recipe features a somewhat quirky ingredient that adds extra richness—duck eggs. The recipe comes from Jill Negronida Hampton, whose son and daughter-in-law, Nico Bryant and Melanie Hook, set up the Bryant Family Farms stand at the market. Their specialty is pastured poultry—chicken, turkey, guinea fowl and duck—and eggs.

Cooking note: If you want to sub in chicken eggs for the duck eggs here, use extra-large ones, or use 2 large eggs plus 1 egg yolk.

Base:

1 cup flour

1/2 cup sugar

1/2 cup (1 stick) cold butter, cut into pieces

Fruit topping:

2 duck eggs

1 cup sugar

1/4 cup flour

1/4 teaspoon salt

1 1/4 cups diced fresh rhubarb

3/4 cup diced fresh strawberries

Garnish (optional): Powdered sugar

Heat oven to 350 degrees. Line an 8-by-8-inch baking pan with a 6- to 7-inch-wide piece of parchment paper that is long enough to extend over opposite sides of the pan. (This is for easy removal of cake after it has baked).

For the base: Combine flour, sugar and butter in a food processor and pulse just until mixture looks crumbly. Transfer mixture to prepared pan and press it firmly and evenly into the bottom. Bake 15 minutes.

Meanwhile, for the topping: Mix eggs, sugar, flour and salt in a bowl until creamy. Fold in rhubarb and strawberries. Pour this mixture over the hot base when it comes out of the oven. Return the pan to the oven and bake until topping is set, 35 to 40 minutes. (If it browns too quickly, cover lightly with foil.) Remove from oven and let cool.

To serve, lift the dessert from the baking pan, remove parchment paper and dust with powdered sugar.

COOL AND CREAMY WILD BLACK RASPBERRY SOUP

4 TO 6 SERVINGS

"Wild black raspberries—also called black caps—grow abundantly on our property, so we used to sell them at the DCFM," says former market vendor Sue Gronholz. "They were a favorite with the chefs, and we always sold out quickly." After reading some recipes, she developed her own version of a Scandinavian chilled berry soup. "Since we also grew several varieties of fresh mint, that found its way into the recipe as well."

Cooking notes:
Cultivated black raspberries, blackberries or red raspberries can easily be substituted for the wild black raspberries in this soup.

Kefir is a low-lactose, fermented milk drink similar to a thin yogurt. Substitute vanilla yogurt thinned with a little water if you can't find kefir.

2 cups wild or cultivated black raspberries, blackberries or raspberries

1/3 cup sugar, or to taste

1 tablespoon lemon juice

1 teaspoon finely grated orange zest

1 teaspoon chopped fresh mint, or to taste

1/4 to 1/2 cup orange juice

2 cups vanilla kefir

Garnish (optional): Additional kefir

Combine black raspberries, sugar, lemon juice and orange zest in a small saucepan. Bring to a boil over medium heat; reduce heat and simmer 15 to 20 minutes, stirring occasionally. (If you like, puree the mixture in a blender or food processor to make it flow better when you press it through the sieve later.) Stir in mint; cover and let cool 20 minutes.

Press the cooled mixture through a fine sieve to remove the seeds. Measure the mixture—there should be 1/2 to 3/4 cup—and add enough orange juice to make 1 cup. (If you get a full cup of berry juice, add a little orange juice anyway, for flavor.) Stir in kefir until thoroughly combined. Refrigerate for several hours before serving.

To serve, portion soup into 4 to 6 small serving bowls and garnish with an additional teaspoonful of kefir, if desired.

Sue Gronholz

MAPLE PUMPKIN HICKORY NUT PIE

6 SERVINGS

Three Native American flavors form a sweet bond in this memorable pie from John and Rosanne Marquardt, of Marquardt's Tree Farm. Being part of a family that's been tapping trees for five generations, John and Rosanne have the maple syrup part well covered, of course. Until recent years, hickory nuts were also readily accessible to them, for their stand was "next door" to former vendors Audrey and Bob Biersach, long-famous at the market for their labor-intensive, compellingly delicious hickory nutmeats. (The pie is also wonderful made with pecans.) The third highlight in the pie is pumpkin. For this, the Marquardts keep it simple and use canned puree, but fresh pumpkin puree is also an option. See page 43 for how to make it from scratch, using small pie pumpkins that are bred for flavor, not size; they show up in droves at the market every autumn.

3 eggs

1 1/2 cups canned or homemade pumpkin puree (see page 43)

1/2 to 1 cup maple syrup (depends on how sweet you like it)

2 tablespoons melted butter

1 teaspoon cinnamon

1 cup chopped hickory nuts or pecans

1 unbaked 9-inch pie shell (see page 228 or use your own favorite recipe)

Garnishes (optional): Hickory nut or pecan halves, sweetened or unsweetened whipped cream

Place a parchment- or foil-covered sheet pan in the oven. Heat oven to 350 degrees.

Use a fork or whisk to beat the eggs in a large bowl until smooth. Add pumpkin, maple syrup, melted butter and cinnamon; stir until smooth. Stir in chopped nuts. Turn mixture into the unbaked pie shell. If using, gently lay hickory nut or pecan halves around outer edge of pie for a garnish.

Place pie on sheet pan in the oven and bake until the crust is golden brown and the filling is set, 45 to 55 minutes. You'll know it's properly set when the center of the filling jiggles slightly when you give the pie a gentle shake—the outer edges should look a little puffy and should not be jiggly at all.

Cool pie completely on a wire rack before serving. Enjoy it straight or with whipped cream.

CONCORD GRAPE PIE

8 SERVINGS

Morren Orchard & Nursery may be best known for their antique apple varieties (see page 232), but there's another elusive, old-timey fruit they bring to market during its brief harvest season in autumn. Concord grapes—aromatic, dark purple and as intensely flavored as jam—were developed from wild species that grew in 19th-century New England and were named after the Massachusetts village where the first of its variety was grown. Grape pie recipes still abound in nearby regions, including New York State, where the bulk of the Concord grape crop now grows.

"Every fall, one or more of our customers will tell us about a grape pie their mother or grandmother used to make," says Lisa Fishman, who owns the orchard with her husband Henry Morren. She shared her recipe, one made with whole grapes, seeds and all, noting that "grape seeds are densely packed with nutrients and are the most 'living,' continuously sustainable form of this ancient fruit." We're all for nutrients, but not so much for dental bills, so we've adapted it here to create a seedless pie.

Cooking notes:

For the pie crust, use the recipe on page 228 or your own favorite recipe.

You'll notice that there are no spices or extracts in this pie. Concords have a well-balanced, wine-kissed taste on their own, and a pie made with them doesn't need extra seasonings or extracts.

5 to 5 1/2 cups Concord grapes (about 2 pounds stemmed grapes)

1 to 2 cups sugar (less for a tarter pie, more for a sweeter one)

3 tablespoons quick-cooking or instant tapioca

2 tablespoons lemon juice

1/4 teaspoon salt

Pinch or two of ground pepper (optional)

Pie dough for a double-crusted 9-inch pie, cool or at room temperature

1 tablespoon cool butter, cut into small bits

Vanilla bean ice cream (optional)

Make the filling: Pop the pulp out of each grape by squeezing it gently into a saucepan (the pulp will come out in one small ball wherever there is a break in the skin). Place the skins in a second saucepan. Bring the pulp to a simmer (don't add any water) over a medium flame; cook until pulp has "melted" into a thick liquid, stirring occasionally, about 5 minutes. Place hot pulp in a fine-mesh strainer that's set over the skins. Press the pulp through the strainer (toss the seeds).

Stir the sugar, tapioca, lemon juice, salt and optional pepper into the skin-pulp mixture. Bring to a simmer and cook, stirring occasionally, to fully dissolve the sugar and soften the tapioca, about 5 minutes. Turn off the heat and cool the filling completely. (It can be refrigerated at this point for later use. Bring it back to room temperature before filling the pie.)

Make the pie: Heat oven to 450 degrees. Line a baking pan with foil and place it in the oven. Working on a floured surface, roll out a little more than half of the pie dough into an 11-inch round. Line a 9-inch pie pan with it. Roll the dough under all around the top of the pie and press the edges together to create a ridge. Scrape pie filling into the pan. Dot the pie with butter pieces.

Roll out the other, slightly smaller portion of the dough into a 10-inch round. Here are two options for the top crust:

Cut the second dough round into strips and make a lattice topping over the filling, saving 2 to 3 of the strips. Lay these around the top edge, over the edges of lattice, then press the dough edges into the formed ridge and flute the ridge all around the pie.

Cut out a 2-inch circle or other shape from the center of the second dough round (this will vent the pie). Lay the dough over the filling, press the top and bottom dough edges together to form a ridge and flute the crust all around.

Place the pie on the foil-lined pan and bake it 12 minutes. Reduce oven heat to 350 degrees and continue baking until crust is light brown and the filling is bubbly, another 25 to 35 minutes. Cool completely before serving.

Lucinda Ranney

"Every fall, one or more of our customers will tell us about a grape pie their mother or grandmother used to make..."

BUTTER AND LARD PIE CRUST

4 OR 5 CRUSTS (FOR 8- OR 9-INCH PIES)

Everyone needs a good pie crust recipe, right? This one comes from veteran pie-maker and DCFM customer Marsha Cannon, who learned about leaf lard at the Rockwell Ridge Farm stand. Leaf lard, rendered from the inner loin area of the animal, is creamier and more neutral-tasting than regular lard. "Combining leaf lard and butter results in a flaky, tasty crust that holds up well," says Marsha. Regular, non-leaf lard will also work very well.

Quality lard is one of those left-behind foods that are gaining attention again at farmers markets; another is flour—fresh-milled, unbleached and organic, and sold by local growers like Meadowlark Organics (see page 230). Try out both in this recipe, which we've adapted from Marsha's, adding details so that even novices can make a crust as good as hers.

Marsha makes enough pie dough for several pies at once, forming it into individual crusts and freezing the ones that won't be used right away. "When I want to make a pie or quiche, I take one crust out of the freezer, remove it from the bag, and let it thaw on the counter while making the filling," she says.

If you don't have enough pie tins to make several crusts at once, you can also form the dough into balls, flatten the balls into disks, wrap them well and freeze them. Then, to make a pie, thaw a dough disk, roll it out and shape it before filling and baking.

4 cups flour

1 1/2 teaspoons salt

1 teaspoon baking powder

1 cup (2 sticks) cold butter, cut into small cubes

3/4 cup leaf or regular lard

2 small eggs

1 tablespoon cider vinegar

1/4 cup cold water

Sift flour, salt and baking powder into a large bowl. Add the butter pieces and lard. Use a pastry blender to cut the fats into the flour mixture until small clumps form when you squeeze a handful of it and a dough begins to form. Don't overwork the dough or it will toughen. You should still see pea-sized pieces of fat in the mixture at this point.

(Alternatively, you can use a food processor to mix the flour, salt and baking powder, then pulse to cut in the fats, scraping down the sides of the work bowl once or twice. Take extra care not to over blend the mixture—food processors work *fast*. Transfer mixture to a large bowl.)

Beat the eggs in a medium bowl with the vinegar and cold water until smooth. Stir this into the flour mixture, blending just enough to moisten the dry ingredients. Again, do not overmix.

Empty the bowl contents onto two or three large, overlapping sheets of plastic wrap. Use the wrap to bring the dough together, then keep pulling the wrap tightly around the dough to form it into a smooth log shape. Cut the log into four or five equal-sized pieces, form each piece into a ball, then flatten it into a thick disk. The disks may be wrapped and frozen at this point, to be thawed and rolled out each time you need one. Or continue with the directions below to form the crusts for immediate use, or to chill or freeze them for later.

Working one at a time and using a floured rolling pin and floured work surface, roll out each disc into a 10- to 11-inch round. Carefully fold the round into quarters, transfer it to an 8- or 9-inch pie pan, unfold it and gently press the dough into and up the sides of the pan. Roll the dough under all around the top of the pie, pressing the edges together to create a ridge. Now flute the edges all around to form an attractive rim.

The crusts will be soft at this point; refrigerate for 30 to 60 minutes to firm them up. Then they can be filled and baked, or wrapped and refrigerated until ready to use. Freeze any that won't be used within a day or two in large, airtight freezer bags.

ON FATS, FLOUR AND PIE CRUSTS

People are growing less afraid of butter and lard these days, thank goodness. We've always known that butter gives the best flavor, and lard makes the flakiest pie crusts. And throughout human history—except for the past seventy years or so, that is—fats were prized above carbohydrates and even protein for a healthy diet. Livestock farmer and DCFM vendor Matt Walter believes we'll look back at the decades when animal fats were reviled and chuckle—or more likely, cringe. "We'll be saying, 'That was a window of time to regret,'" he predicts.

Just as we're relearning that moderate consumption of the fats from humanely, sustainably raised animals can be healthful, so it goes with flour—another denatured, overindustrialized foundation ingredient that's finding its way back into our good graces. As more regional growers, such as the DCFM's Meadowlark Organics, supply high-quality, stoneground flours to farmers markets and food stores, we're seeing again how flour—the base for so many of the world's most time-honored foods—can be fresh, full-flavored and yes, good for us.

Terese Allen

This is all especially good news for pie lovers—and man, but there are a lot of those on this planet. Sweet or savory, open-face or double-crusted, hand-held or sliced into wedges, pies are beloved in countless cuisines around the globe. Moreover, as every baker knows, a great pie starts with a great crust—and pie crust is all about flour and fat. We offer a recipe on page 228. Try it out with first-rate ingredients from local growers in your next *quiche Lorraine*, Cornish pasty or pumpkin pie.

This is all especially good news for pie lovers—and man, but there are a lot of those on this planet.

MORREN ORCHARD BAKED APPLES WITH CIDER SAUCE

6 SERVINGS

Henry Morren and Lisa Fishman break into smiles when you mention this dish. That's because they associate it with their 30th wedding anniversary trip to Washington Island (on Lake Michigan off the tip of Wisconsin's Door County peninsula), where they came across a recipe for apple cider sauce in the kitchen of a tiny fisherman's-cottage-turned-museum.

The couple has a thing for old-fashioned foods. At Morren Orchard and Nursery, they grow over 100 varieties of apples, including many antique and heirloom American, French and English varieties. They also fresh-press apples for cider, and raise pears, plums and other fruits and vegetables.

Baking apples is the way "to enjoy apples in their purest form—apart from eating them fresh," says Lisa. The sauce is optional, she adds, but very much recommended as a delicious "puddle" for the apples to bask in at serving time. It can also go inside the apples before baking them.

The Morren Orchard apple varieties she recommends for the dish include:

- Cox's Orange Pippen (ripens mid-September): Aromatic, multi-dimensional flavor, orange-red coloring.
- Brown's Apple (late September): Tangy, scented, dark red, clean flavor.
- Rhode Island Greening (early to mid-October): Official apple of Rhode Island, tart and juicy, greenish-yellow, similar to Granny Smith.
- Roxbury Russet (mid-October): Traces to colonial times, crisp, spicy-sweet-tart, russet skin

Cider sauce:

1/4 to 1/2 cup sugar

3 tablespoons cornstarch

1/4 teaspoon salt

1 cup fresh-pressed apple cider

2 tablespoons lemon juice

2 tablespoons butter (or vegan substitute)

Apples:

6 firm, sweet-tart baking apples such as those listed on previous page

1/2 cup chopped walnuts or almonds

1/2 cup raisins, dried cranberries or minced crystallized ginger

1/4 cup brown sugar, maple syrup, honey or cider sauce (see opposite page)

1 1/4 teaspoons cinnamon or apple pie spice

Garnish (optional): Sweetened or unsweetened whipped cream

Make the cider sauce: Combine sugar, cornstarch and salt in a saucepan. Mix in the cider, bring to a simmer over low heat, stirring constantly, and cook until mixture is golden and thickened, 5 to 6 minutes. Remove from heat and stir in lemon juice and butter.

For the apples: Heat oven to 375 degrees. Use a paring knife to cut out a 1 1/2-inch-wide opening into the top center of each apple. Use a melon baller or grapefruit spoon to core each apple, taking care not to pierce the bottom. Using the knife again, peel a 1-inch-wide strip of skin off each apple all around its top. Now, starting at the top of the skin, make five vertical 1-inch slits into the skin around each apple. (This will prevent the apples from exploding as they bake, which occurs when the heating fruit expands and splits open the skin.) If the apples have "bumpy" bottoms, shave a very thin slice off the blossom end, taking care not to pierce through the bottom, so that they'll sit flat in the baking dish.

Combine the nuts, dried fruit or ginger, and the sugar, syrup, honey or cider sauce, adjusting the proportions as you like, or even choosing other, similar ingredients from your pantry. Select a baking dish that'll hold the apples with just a little space between them, and set the apples inside it. Fill the cores with the filling.

Bake until the apples are just tender when pierced with a fork, about 45 to 55 minutes. Check them once or twice in the final quarter of the cooking time; apples vary a lot, and you don't want them to get mushy.

Cool apples 10 to 15 minutes before serving them. Rewarm the cider sauce, stirring it, and ladle some onto each of 6 serving dishes. Place a baked apple atop the sauce. Whipped cream would not be out of order here!

APPENDIX A

THE FIRST FIFTY YEARS: AN INFORMAL HISTORY OF THE DCFM

EARLY HUMAN HISTORY TO 1800S: Open markets characterize humankind's earliest organization into community. Their growth goes hand in hand with the development of civilization. In colonial times, fairs operate throughout our country and are typically held in town squares, where commerce and social connections have long mingled. In pre- and early Wisconsin days, native peoples barter with fur traders and early settlers. The recording of established markets in the state begins in the mid-1800s.

MID-1800S TO EARLY 1900S: An urban farmers market is held in various locations near the Capitol Square in Madison, including at the City Market, built in 1910 but located too far from the Square for lasting success. It closes after World War I. In ensuing decades, family farm stands operate near the city limits and some small farm markets in nearby communities exist, but there is no large, central market. In the mid-20th century, as more large grocery stores and suburban malls open, many municipalities close their outdoor markets. Farmers markets don't begin to return until the 1970s and 80s.

1971 TO 1972: Spurred by a burgeoning back-to-the land movement and increasing consumer interest in fresh produce, the idea of (re)establishing a farmers market in Madison takes root. Initial efforts by Dane County Extension agents James Schroeder and Ron Jensen to found local markets meet with resistance from the shopping malls they approach as potential locations. ("Too messy," said mall managers.)

Madison Mayor Bill Dyke, aware of the economic and cultural benefits of farmers markets, gathers key players to make one happen in Madison. Grower (and future Dane County Executive) Jonathan Barry—who, among other farmers, wants to see a central market happen—is recruited to manage the new market. He and Jensen further investigate possible

sites, but nobody bites. Mayor Dyke favors a Capitol Square location and enlists Governor Lucy to convince Capitol staff that it would be in the interest of the State. Final hurdles are surmounted when the Central Madison Committee—whose mission is downtown revitalization—agrees to sponsor the market by covering trash costs and including the market in its liability insurance. Dane County UW-Extension also comes on board as a sponsor.

The first market opens on Saturday, September 30, 1972 on a Capitol Square inlet, with less than a dozen vendors, who are quickly besieged by produce-hungry shoppers. The following Saturday two or more thousand customers descend on 85 vendors, with more jumps in crowd size and the number of vendors during the final two markets that year. Whatever doubts of the market's potential that remain are put to rest.

1972 TO 1976: Jonathan Barry's diligent management and promotion of the market are central to its early success. So are such strategies as limiting the market to agricultural products, requiring a fee from vendors, and creating a growers council to represent the interests of the vendors. Attendee and vendor numbers continue to rise. The market's season is lengthened; early-season crops and bedding plants make their first appearance, increasing the DCFM's allure and usefulness to both growers and buyers. A midweek market, held on Wednesdays, is added to the schedule. In these first years of the market, it is largely the only place for area residents to buy farm-fresh products.

1976 TO 1979: Melon growers Judy and Dan Peterson take the reins as co-managers. The Growers' Council gets its sea legs as an advisory group for market sponsors and managers, and begins to evolve into a governing body. The Wednesday market matures—settling into its primary role as a mid-morning to mid-afternoon venue for downtown workers—especially those toiling away in nearby government buildings.

Odessa Piper, an early DCFM bakery vendor and the chef-proprietor of the fledgling (and eventually nationally renowned) L'Etoile restaurant, begins basing her menus on ingredients purchased from other vendors. Other area chefs also buy produce at the market. Over time, Piper's influence becomes a dynamic factor in the market's success, and her culinary model evolves into a regional movement of farm-to-table dining.

The Petersons publish market rules and policies and lay the groundwork for obtaining not-for-profit status. They establish stall locations for regular vendors, ending the need for these farmers (and their families) to hold a place in line by sleeping in their trucks overnight. The seasonal stall assignments are determined by lottery; later, a seniority-based lottery makes the assignments permanent. Now seasonal vendors can sell at the same location from year to year; likewise customers can know where to find them. The market keeps right on growing.

1980 TO 1983: New co-managers Jo and Paul Prust, who are former DCFM egg vendors, hit the ground running. Their work with attorney Sarah O'Brien results in the DCFM formally becoming a not-for-profit organization. It creates clearer definitions for membership, allowable products and how to handle violations. Vendors, for instance, are now required to participate in the production of their foodstuffs (no buying and re-selling). The ever-increasing size and scope of the market parallels moves towards self-government. The market's sponsors appoint a board of directors, and its Growers' Council becomes an elected (but still advisory) group. Market overseers grapple with—and find some solutions for—food sampling issues, inspections of vendors and people traffic on the Capitol lawns. The DCFM is growing up.

1984 TO 1989: Peak-season Saturday markets now host about 150 vendors and see crowds of 10,000 attendees. Mary Walters, the first manager who is not a vendor, takes the reins of the organization. A former meat grader, her education and experience serve her well for the health and safety considerations that have become necessary with the increase of many value-added foods at the market. Jams, relishes, vinegars, pesto, pasta, plus fresh trout, frozen meat and other goods, must now satisfy rigorous checks. Elected members of the Growers' Council replace the appointed board of directors (who become advisors), and the DCFM officially becomes independent from its original sponsors.

The board crafts a mission statement that emphasizes: support of Wisconsin growers, crops and products; the health and welfare of eaters; agricultural and culinary education; and urban-rural connections. Two of Walters' brainchildren begin their annual run—the Market Dinner, a social and culinary showcase for the market community, and the indoor Holiday Markets, which allow sales of agriculturally related crafts (and still operate today, at Monona Terrace Community and Convention Center). Controversy comes to the DCFM when concerns about

traffic congestion, litter, lawn trampling and the use of Capitol facilities threaten the market's location on the Square.

LATE 1989 TO 2000: Grower and educator Mary Carpenter begins her run as market manager, with support from her husband Quentin. The first major hurdle is to see the market through the Capitol Square congestion controversy. After a series of efforts to resolve the issue, solutions proposed by the City of Madison—the suspension of bus traffic during market hours, trash control measures, and the assignment of Capital police to market security—do the trick. The DCFM is back on track (and going at top speed, as ever).

The Carpenters redevelop the membership seniority system to include all vendors—that is, both seasonal and by-the-week sellers. They initiate the first thorough market survey. The DCFM board hones vendor requirements and rules, and inspections are stepped up. The sellers' roster hits nearly 400; a membership cap, a waiting list for new vendors, a monthly vendor newsletter, and board meetings open to all vendors ensue. An educationally focused DCFM website is launched. The Saturday market begins hosting weekly cooking demos and workshops. The market's right to create and enforce its own rules is confirmed when a lawsuit brought by an expelled vendor is judged by the courts to be frivolous—but not before the Carpenters must spend months dealing with it. They also institute a rainy day fund for the organization.

During these years, national and regional media gets on the DCFM bandwagon, giving the market much attention as among the best in the nation. Mary Carpenter consults with several cities looking to start or improve their own markets. The DCFM features prominently in *Fresh Market Wisconsin*, by Terese Allen, the first state-based, farmers' market-focused cookbook. The market's local impact broadens when two volunteer-based endeavors begin collecting surplus produce for food pantries. Hmong farming families making a new life in Wisconsin diversify the market and enrich the region's culinary culture. (For more info, see page 146). Organic produce, heirloom and hydroponic vegetables, fresh flowers, grass-fed meats and an ever-growing array of new types of crops and products increase in popularity. The DCFM matures into a force—locally, regionally and nationally.

LATE 1989 TO 2000

2001 TO EARLY 2003: Bill Warner and Judy Hageman, husband-and-wife DCFM vendors known for their early produce, begin their tenure as co-managers. In September 2001 (amidst the 9/11 terrorist attack crisis), they announce that the market will become year-round. The indoor Holiday Markets, which end in late December, are now followed by Winter Markets held from January to April, when the outdoor Capitol Square venue commences each year.

Hageman and Warner initiate an info booth at the Winter Market and establish a weekly Market Breakfast, a community social event that highlights local crops and products, while teaching shoppers how to utilize the cold-season bounty. Area restaurateurs showcase their skills at the breakfast, and the relationship between local chefs and DCFM farmers flourishes. REAP Food Group's Food for Thought Festivals, held alongside the Saturday market, hit their stride as an annual event that goes hand in hand with the DCFM's mission to educate people about local foods and sustainability. New in 2003 is the Friends of the Dane County Farmers' Market, a volunteer organization established to support educational and charitable initiatives, including cooking and growing demonstrations, youth activities, REAP's Farm Fresh Atlas (a guide to regional food sources), food pantry donations, and free transportation for seniors to the market.

The State of Wisconsin launches FoodShare programs that allow low-income people to buy farmers market produce with vouchers, and the DCFM makes them available to its shoppers.

The Dane County Farmers' Market is now the largest farmers market in the country selling only regionally grown crops and requiring producers' attendance. Numerous smaller satellite markets are in operation throughout Madison and nearby communities. Eventually more than 30 markets are held weekly in the area.

2003 TO 2014: Geologist and flower grower Larry Johnson becomes the DCFM's first full-time manager. Early in his tenure, a market survey shows $8 million in DCFM vendor sales, with $6 million generated for downtown Madison merchants and restaurants. Ready-to-eat, frozen or canned foods made with regional ingredients hit the market scene, including ravioli, pasties, soups, quiche-lets and pizza.

In the early 2000s there is much national coverage about the market, including articles in the New York Times, Huffington Post and National Geographic magazine. Local media run columns featuring market vendors, local crops and products, and recipes. *The Dane County Farmers'*

Market: A Personal History, by former DCFM manager Mary Carpenter (with Quentin Carpenter) is published. The market's historical records are processed and archived at the Wisconsin Historical Society.

Odessa Piper sells L'Etoile to the restaurant's chef de cuisine, Tory Miller, and his sister Traci Miller in 2005. Tory, along with a growing number of area chefs, makes his own mark at the DCFM, and furthers the goal of a regionally reliant food system. His community contributions include volunteer guest-chef gigs at the Market Breakfasts and other area events, and teaching "local foods" culinary education classes in Madison schools. (In 2012, Miller—like Piper before him—is named Best Chef of the Midwest by the James Beard Foundation.)

The influence of area chefs on the evolution of the DCFM, and on the local foods movement, is now significant. More restaurants are listing suppliers on their menus, which helps drive everyday buyers to the market. Chefs request specialty produce and products from vendors, and vendors respond by growing or producing them. And vendors, in turn, offer new crops and products that motivate chefs to use them. The diversity of foodstuffs and buyers continues to grow, a dynamic that keeps the DCFM at the forefront of markets nationwide.

In 2008 the DCFM weathers a tanking national economy and sidewalk construction on the Square. The Friends of the DCFM help bring equipment to the market that allows food stamp recipients to purchase items with state-issued debit cards. By 2010, there is a vendor waiting list of five years. The system is modified to better manage membership and stall availability. Average attendance is 18,000 per week and the top turnout to date is 25,000. In 2011, southern Wisconsin boasts 45 farmers markets. Statewide the total is 231—only California and New York have more.

The DCFM's website develops into a useful resource for growers and customers; it features vendor highlights, recipes and seasonal product lists. The market's presence on social media also expands. Indirectly connected to the DCFM are other notable initiatives that occur during this time: the formation of the Dane County Food Council; local and state-wide "eat local" challenges, meant to shine a light on the pleasures and benefits of local foods; Local Foods Summits, annual conferences held around the state; the growth of community supported agriculture (CSA) and the publication of a new edition of *From Asparagus to Zucchini: A Guide to Cooking Farm-Fresh Seasonal Produce*, by FairShare CSA Coalition; and a new online vehicle for ordering local produce for pick-up at the DCFM.

2014 TO 2016: The new manager is Bill Lubing, a writer-photographer specializing in local foods who has been editor of the DCFM newsletter for six years. Lubing's goal is to help more shoppers view the DCFM as a viable place to do their overall food shopping. "Nothing unusual, just: This is where we get our groceries," says Lubing, as quoted in a local newspaper article. Lubing coordinates various DCFM public events, upgrades and streamlines recordkeeping, and manages market disruptions due to ongoing construction on the Square. He works with a committee to convert member archives to electronic form, initiates canning workshops sponsored by the DCFM and, with University of Wisconsin staff, coordinates a comprehensive survey to measure patron attitudes and behaviors regarding the market. He also hires the market's first assistant manager.

During this period, proposals for two major urban projects are under discussion. One is the Madison Public Market, an inclusive year-round indoor marketplace showcasing locally grown and prepared foods, scheduled to debut in 2025. The other is the renovation of the city's historic Garver Feed Mill into a local foods production center and events venue. Garver Feed eventually becomes the location for the DCFM's Winter Market.

2016 TO 2021: Food systems planner Sarah Elliott, who has led local foods initiatives for non-profit organizations and the State (including Wisconsin's Farm to School Program), takes charge. During her early years as manager, Elliott: hires assistant manager, Jill Groendyk; facilitates the DCFM's rebranding with fresh logos, a new tagline ("Harvesting Wisconsin Goodness") and a revamped website; pivots to a third-party inspection system for vendors; and streamlines market operations and financials to develop a rainy day fund (which would all too soon prove crucial to the market's ability to withstand the COVID-19 pandemic).

Two intentional moves to larger venues transpire during Elliott's tenure. The Winter Market transfers from its long-cramped space at the downtown Senior Center to the city's new community jewel, Garver Feed Mill, on the city's east side. (It is so quickly successful that the DCFM must immediately add parking attendants and additional nearby parking.) The other planned change-up is the DCFM's once-a-year relocation during Madison's annual Art Fair on the Square to Breese Stevens Field, also on the east side. Both moves helped reinvigorate these markets and connect new customers to DCFM members.

Noteworthy during this time are efforts to deepen public understanding of the market's purpose and potential. Elliott and Groendyk foster a view of the market as an entity beyond "tourist attraction," a place where the community gathers, does its weekly grocery shopping and sustains a direct connection with the people who grow, raise and make their food. Local newspapers publish articles about the increase in food choices at the market over the years, the diversity of its customers, farmers and producers, and the growing number of second- and third-generation vendors. The DCFM also becomes a partner of the new Madison Night Markets, a new-style urban experience cropping up around the country that mixes local foods, music, arts and activities.

The most challenging period in DCFM history begins in early 2020, as the world reels during the onset of the COVID-19 pandemic. The market launches a drive-thru outdoor market that uses on-line pre-ordering and operates in shifts. It is held first in the Garver Feed Mill parking lot, and then at the Alliant Energy Center, a large-scale events complex in town. These unforeseen venue changes and the outbreak's social restrictions are a major blow to the DCFM—attendance, local foods access, the livelihood of members and the market's beloved personal connections all suffer. There's no food sampling, no music, no community non-profit tables, no bathrooms. Masks, social distancing and a strict safe shopper code of conduct are now the order of the day.

The DCFM works with public officials to develop a safe, efficient model that continues to connect customers and members, but also meets ever-changing public health requirements throughout the crisis. Within a few months, the market is able to add a walk-up shopping option to the operations. It partners with FairShare CSA Coalition to launch an Emergency Farmer Fund, which quickly raises nearly $50,000 to help the organizations' members navigate pandemic disruptions. The market also maintains Food Share benefits throughout the upheaval.

The DCFM remains at the Alliant Center until the early summer of 2021. (During the winter it features indoor, drive-thru shopping at one of the complex's livestock pavilions.) Around the region, some growers and producers switch to home delivery, neighborhood pick-ups, CSA arrangements and other business models to stay afloat. Some retire altogether. Still, like the DCFM, most farmers markets in the region remain open, maintaining critical sales opportunities for local producers, and even strengthening the farm-to-table vendor-customer relationship, as both sides grieve the loss of more direct contact.

The pandemic alters the market's course and its history. But, along with 9,000 other farmers markets across the county (including about 300 in Wisconsin), the DCFM helps fill gaps in the food chain caused by COVID-19 shutdowns, offers eaters an alternative to shopping indoors at grocery stores, and responds to a new surge of interest in local foods.

Elliott, who has been praised by the director of the Wisconsin Farmers Market Association for the "amazing job" she did during the market's darkest hours, retires from the position in April of 2021. A few months later, the market returns to the Capitol Square.

2021 TO PRESENT: Jamie Bugel, a plant scientist and the former leader of Madison's Eastside Farmers' Market, becomes the DCFM's manager, with Jill Groendyk serving as co-manager for a short time. By June of 2021, with COVID-19 restrictions loosening, the market is back in business on the Capitol Square. At the first one, some 100 booths set up and about 20,000 eager shoppers attend. The relief and joy are palpable. "This is very important to me because it's my living," says vendor Gretchen Kruse in a Wisconsin State Journal piece. "It means we have some semblance of order again." The same article quotes longtime pesto producer Mark Olson: "This is home to me. This is my family. This is my community."

Bugel and assistant manager Rachel Figueroa work to bring normalcy back to the market and also to move it forward. They create a position for a Food Access Coordinator, making the management of government supplemental programs in-house for the first time. They return the Winter Market to Garver Feed Mill (with indoor COVID-19 guidelines in place) and expand it to 40 vendors. They win a grant to fund free compost drop-off at the Wednesday market.

In 2022, market attendance is back to pre-COVID-19 numbers and there are 35 vendors on the membership waiting list. All pre-COVID-19 DCFM locations are operating: outdoors on Saturdays at the Capitol Square; outdoors just off the Square on Martin Luther King Boulevard; indoors at the Monona Convention Center for the Holiday Markets; and indoors at Garver Feed Mill for the Winter Markets.

2022 also marks the DCFM's 50th anniversary, with special committees managing events and initiatives to celebrate the milestone. Among these are a fundraiser for DCFM programs that features decorated market wagons, a public picnic on the Capitol lawn, and the creation of *The Dane County Farmers' Market Cookbook: Local Foods, Global Flavors*, published in 2023. Happy birthday to the DCFM ... and here's to the next fifty years.

2021 TO PRESENT

Timeline Bibliography

Allen, Terese. *Fresh Market Wisconsin: Recipes, Resources and Stories Celebrating Wisconsin Farm Markets and Roadside Stands.* Amherst Press, 1993.

—*Wisconsin Local Foods Journal: Sustainable Eating All Through the Year,* co-authored by Joan Peterson. Gingko Press, 2012.

—The *Flavor of Wisconsin: An Informal History of Food and Eating in the Badger State.* Revised and Expanded Edition, co-authored by Harva Hachten. Wisconsin Historical Society Press, 2009.

—Isthmus, "Local Flavor" and "Farmers Market" newspaper columns, 1999–2011.

Carpenter, Mary with Quentin Carpenter. *The Dane County Farmers' Market: A Personal History.* University of Wisconsin Press, 2003.

Carpenter, Quentin. Script for Dane County Farmers' Market's 50th anniversary celebration, Capitol Square, October 1, 2022.

—Email correspondence with author, February, 2023.

Dane County Farmers' Market website. "Start of the Dane County Farmers' Market." Dcfm.org, undated.

Eisen, Mark. "What's Next for the Dane County Farmers' Market?" Wisconsin Examiner.com, July 6, 2022.

Karon, Sarah. "History of the Farmers' Market." Channel3000.com. April 27, 2012.

Piper, Odessa. "Odessa Piper Reminisces: DCFM's 50th Anniversary," Edible Madison, November 14, 2022.

Smith, Susan Lambert. "A longtime vendor reflects on four decades at the Dane County Farmers' Market," Isthmus, July 9, 2022.

Wagener, Candice. "Dane County Farmers' Market Celebrated 50 Years." Edible Madison, September 7, 2022.

Wisconsin State Journal and Capital Times archives, 1970-present.

Also:

Interviews with former DCFM managers Bill Warner, Larry Johnson, Bill Lubing and Sarah Elliott, and with current manager Jamie Bugel, via email, February, 2023.

APPENDIX B

WHAT'S AT THE MARKET: LIST OF CROPS AND PRODUCTS

While this is an extensive listing of items sold at the Dane County Farmers' Market, it's not a comprehensive one. The market has a changing beauty—many foodstuffs are available week in, week out, while some come and go based on season, supply level and vendor attendance. Many items listed below come in a huge array of flavors, forms or varietals. New crops and products show up every year, and occasionally something becomes unavailable when a producer retires or changes offerings. Base your meals on what's available at the market, what looks good to you and what makes your mouth water. Buy according to what you value and can afford. You won't be disappointed.

Bakery

Bagels
Bars
Biscotti
Bread (wheat, rye, multi-grain, etc.)
Bread, cheese
Bread, gluten-free
Bread, semolina
Bread, sourdough
Bread, whole grain
Brownies
Candies
Cinnamon rolls
Cookies
Crackers
Cream puffs
Croissants
Danish
Doughnuts
Empanadas
Focaccia
Fritters
Gluten-free breads and pastries
Graham crackers
Granola
Keto/paleo breads and pastries
Muffins
Pastries
Panettone
Pies
Pies, hand
Quick breads
Scones
Tarts
Tortes
Tortillas
Vegan/vegetarian breads and pastries

Cheese and Dairy

Artisan or specialty cheese
Asiago
Baby Swiss
Baked cheese (also called bread cheese)
Blue
Brick
Brie
Butter
Butterkäse
Camembert
Cheddar
Cheddar, aged
Cheddar, bandaged
Cheese curds
Chevre
Colby
Cream cheese spreads
Farmers cheese
Feta
Flavored cheeses
Goat butter
Goat cheese
Gorgonzola
Gouda
Gruyere
Havarti
Monterey Jack
Muenster
Parmesan
Pepper Jack
Romano
Sheep cheese
Smoked cheese
Spreads
Swiss
Yogurt drinks

Fruits and Nuts

Apple butter
Apple cider
Apples
Apricots
Aronia
Bitter Melon
Blackberries
Black raspberries
Black walnuts
Blueberries
Butternuts
Cantaloupe
Cherries
Cranberries (fresh, dried)
Currants (black, red, white)
Elderberries
Frozen fruits
Gooseberries
Grape juice
Grapes, Concord
Ground cherries
Hazelnuts
Hickory nuts
Honeydew
Lemons
Melons

Muskmelon
Nectarines
Peaches
Pear butter
Pear cider
Pears
Plums
Raspberries
Rhubarb
Strawberries
Watermelon

Grains and Flours

All-purpose white flour
Bread flour
Buckwheat
Cornmeal
Heritage wheat flour
Oats
Pancake mix
Pastry flour
Polenta
Rye berries
Rye flour
Spelt flour
Wheat berries
Wheatgrass
White corn grits
Whole wheat flour

Herbs and Aromatics

Anise hyssop
Basil
Bay leaves
Bee balm
Black garlic
Borage
Chervil
Chiles, fresh and dried
Chives
Cilantro
Curry leaves
Dill
Dried herbs
Elderflowers
Epazote
Garlic
Ginger
Ginseng
Lavender
Juniper berries
Kafir limes and leaves
Lemon balm
Lemon verbena
Lemongrass, fresh and dried
Lovage
Marjoram
Mints
Oregano
Parsley
Pipicha
Rosemary
Sage
Sumac
Summer savory
Tarragon
Thyme
Turmeric

Leafy Greens

Arugula
Asian greens
Broccoli rabe
Chard, Swiss and rainbow
Collards
Dandelion greens
Escarole
Frisée
Kales
Lettuce, butter
Lettuce, head
Lettuce, iceberg
Lettuce, leaf
Lettuce, romaine
Mâche
Mesclun
Microgreens
Mizuna
Mustard greens
Purslane
Radicchio
Salad mix
Sorrel
Spinach
Turnip greens
Watercress

Meat, Poultry and Eggs

Beef
Beef, grassfed
Bison
Bratwurst
Chicken
Chicken, pastured
Chorizo, Mexican-style
Dog bones
Duck
Eggs, chicken
Eggs, duck
Eggs, goose
Eggs, quail
Elk
Goat meat
Goose
Guinea fowl
Ham
Hot dogs
Italian sausage
Jerky
Lamb
Lamb, pastured
Landjaeger
Lard
Liver sausage
Meatloaf
Offal
Pork
Pork, pastured
Rabbit
Salami
Sausage, many varieties
Snack sticks
Soup bones
Steak
Summer sausage
Turkey
Venison

Mushrooms

Chaga
Chanterelles
Chestnut
Cremini
Enoki
Hen-of-the-woods
Lion's mane
Maitakes
Mushrooms, dried
Mushroom kits
Morels, fresh or dried
Oyster mushrooms
Portobello
Reishi
Shiitake
Turkey tail
White button
Wild mushrooms

Non-Edibles

Candles, beeswax
Catnip
Compost
Flowers (cut, dried, bouquets, plants)
Hides
Gourds
Holiday decorations
Leather
Ornamental corn
Ornamentals
Potting mix
Sheepskins
Smudge sticks
Soap
Wood chips
Wool, raw
Wreaths
Yarn

Plants

Annuals
Berry bushes
Ferns
Flowers
Fruits

Fruit trees
Hanging baskets
Herbs
House plants
Nursery stock
Perennials
Succulents
Vegetables

Products and Packaged Foods

BBQ rubs
Black garlic
Bloody Mary mix
Brownie mix
Candied chiles
CBD products
Chile, candied
Chile seasonings
Chile sauces
Chiles, dried
Chocolates
Chutney
Confections
Dips
Dog food and chews
Dried fruits, vegetables, herbs, etc.
Enchiladas
Flax seed
Flax seed oil
Fermented foods
Fruit butters
Gift boxes
Gourds
Harissa paste
Hemp products
Hominy, dried
Honey
Honey, whipped
Honeycomb
Hops
Hot sauces
Hummus
Jams
Jellies
Kimchi
Lavender
Lasagne
Maple candy
Maple cream
Maple sugar
Maple syrup
Pancake mix
Pasta
Pesto
Pickled mushrooms
Pickled vegetables
Pickles
Popcorn
Ristras
Salad dressings
Salsa
Salsa kits
Sauces
Sauerkraut
Sorghum
Soups
Stocks
Sunflower oil
Tamales
Teas
Tomato sauce
Tortillas
Vegetable noodles
Vinaigrettes
Vinegars
Wild rice

Vegetables (see also Leafy Greens, Mushrooms)

Artichokes
Asparagus
Asparagus beans
Bean sprouts
Beans, fresh, dried and shelled
Beets
Bitter melon
Bok choy
Broccoli
Broccolini
Brussels sprouts
Burdock root
Cabbage (green, purple, red, etc.)
Cardoons
Carrots (orange, purple, yellow, etc.)
Cauliflower (white, green, orange, purple)
Celeriac
Celery
Chiles, fresh and dried
Chinese cabbage
Corn, sweet (fresh, frozen, dried)
Cranberry beans (shell, dried)
Cucumbers
Cucumbers, hydroponic
Daylily buds (fresh, dried)
Edamame
Edible flowers
Eggplant
Endive
Fava beans
Fennel
Fiddlehead fern tops
Frozen vegetables
Garlic (fresh, dried, spring)
Garlic braids
Garlic scapes
Ginger
Green onions
Jerusalem artichokes
Jicama
Kohlrabi
Leeks
Long beans
Mouse melons
Napa cabbage
Nettles
Okra
Onions
Onions, green
Onions, spring
Parsnips
Pea shoots
Peanuts
Peas, shelled
Peas, snap
Peppers, hot
Peppers, sweet
Popcorn
Potatoes (white, red, purple, yellow, etc.)
Pumpkins
Radishes
Radishes, daikon
Radishes, watermelon
Ramps
Rhubarb
Romanesco
Rutabaga
Salsa kits
Shallots
Soybeans
Spinach
Spinach, winter hoophouse
Spring garlic
Sprouts
Squash, summer
Squash, winter
Squash blossoms
Squash vines
Sweet potatoes
Thai eggplant
Tomatillos
Tomatoes
Tomatoes, hydroponic
Turnips
Turnips, salad
Vegetable noodles
Water spinach
Zucchini

ACKNOWLEDGMENTS

My thanks go first and foremost to Dane County Farmers' Market vendors Joan and Ted Ballweg. This book is their brainchild. I'm more or less its surrogate mother—the person they asked to produce (and co-raise) the kid. Our co-parenting project began very early in 2022 as an initiative of the DCFM's 50th anniversary year. Joan and Ted believed it was the perfect time—as well as high time—for what is arguably the country's best farmers market to have its own cookbook. They asked me what I thought and wouldn't I like to write it. Before I knew it, we were developing a concept, soliciting recipes, rounding up photographers and recipe testers, and securing a publisher. From day one onward, the Ballwegs and I have been a team. I know it's a cliché, but I'm going to say it anyway: I couldn't, nor wouldn't, have done it without them. Thank you, Joan and Ted, for the vision and values we've shared, for your multi-layered skills and labors, your steadiness and humor, and your friendship. Also, the champagne.

I'm deeply grateful for and to these key players: Chef-restaurateur-community leader Tory Miller, who not only wrote the Foreword (is there anything this man cannot do?), but also said "no problem" instantly, cheerfully, every time we asked him for yet another favor. Kristin Mitchell of Little Creek Press, our intrepid, extraordinarily capable and always-reassuring publisher, and the talent who designed this gorgeous keepsake. Former DCFM manager Bill Lubing, the book's photo editor—aka picture wrangler and primary photographer—a generous, good-natured man whose involvement in the project brought back fond memories of past collaborations and has been a special treat for me. Bill also serves as our PR and sales manager and we're extremely lucky to work with him in that capacity, too. One more: the book's copy editor Wendy Allen, no relation, though I'd be proud if she was because she's the most vigilant and finickiest (read: best damn) proofer a writer—or reader—could ask for.

I'm beholden to several groups of people for their contributions, and I'll start with the recipe testers: Mimi Dane, Deb Shapiro, David Flesch, Sarah Brooks, Pamilyn Hatfield, Jackie Hass, Ruth Bronston, Edith Thayer, Char Thompson, Rachel Figueora, Carole Blemker and Megan Bjella. I can't tell you how fun it was deliberating ingredients, measurements, methods and flavor with these cooking geeks, and how much their efforts have enhanced this cookbook.

Then there's the photography team, whose work makes the recipes and the market come alive so beautifully: Bill Lubing, Lucinda Ranney, Lois Bergerson and Joan Ballweg, plus a goodly number of recipe contributors and testers who offered shots of their creations. And the DCFM gang, market leaders who supported the project in various ways: manager Jamie Bugel, assistant manager Rachel Figueora, former board president Cliff Gonyer, current president Henry Morren and all the DCFM board directors. And of course, the recipe providers, whose delicious efforts are at the core of this book. I thank not only the cooks whose contributions became a part of the collection, but everyone who submitted their beloved recipes and warm-hearted stories—all you DCFM vendors, shoppers and supporters who have proven to me all over again that food is love.

Now for some grateful shout-outs to: individuals who assisted with the history section (former DCFM managers Larry Johnson, Bill Warner, Bill Lubing, Mary and Quentin Carpenter and Sarah Elliott, and current manager Jamie Bugel); fellow members of my longtime Books for Cooks reading group, who let me bounce ideas off the walls of Zoom and pick their brains about concept, content and recipe editing; publications that gave the book a jump-start by running articles during the recipe-collection phase *(Capital Times, Isthmus, Edible Madison, Up North News* and *Wisconsin Examiner);* readers who reviewed parts or all of the book (Jeannie Kokes, Bill Lubing, Lois Bergerson and the Ballwegs); the kind luminaries who wrote book blurbs; all the current and future volunteers who help the DCFM get this cookbook into your hands; and the miscellaneous few who gave extra support in ways large and small: Judy Fisher, Mark Allen, Barb Pratzel, Jeannie Kokes, Mary Bergin, Kristel McHugh, Heidi Gilbertson, Kristin Groth, Billie Greenwood and my wise, wonderful, forgiving and tall husband, Jim/JB Block.

I wish I could name and thank every vendor and shopper friend I have made at the Dane County Farmers' Market over the past four decades. At first, I tried to list them, but then I got a little weepy ... and then I got a little worried that I'd miss somebody. Truth is, there isn't enough room to give those rich relationships justice, anyway. I could tell you a thousand people stories—about the garrulous, charismatic long-retired meat vendor I still miss seeing each week, or the shopper-rogue who steals my cart when I'm not looking, or the cheesemaking couple I "argue" with over money (they try to undercharge me and I try to overpay them). There's the sometimes jubilant, sometimes woeful Packer gab sessions I've shared for years with a grower buddy (and with her father, before he passed away). The joke-of-the-day exchanges with a fun-loving seller that launched a now-deep friendship.

Not long ago I read a piece in the *New York Times* about the casual connections we make at regularly visited places, like farmers markets, dog parks and bus stops. Paula Span wrote, "Psychologists and sociologists call these sorts of connections 'weak ties' or 'peripheral ties,' in contrast to close ties to family members and intimate friends. ... Such seemingly trivial interactions have been shown to boost people's positive moods and reduce their odds of depressed moods." Span quoted a psychologist who said that the sense of belonging that weak ties confer is "essential to thriving, feeling connected to other people." Nothing weak about that, right?

So, market friends of mine, I trust you know who you are. Or maybe you don't, so here it is: You're the main reason I wrote this book. I bet you thought it was all about the food.

INDEX

Italicized page numbers indicate photos of the named recipe or person.

M

N

O

P

Q

R

S

T

U

V

W

Y

Z

ABOUT THE AUTHOR

Jeannie Kokes

Terese Allen has been called Wisconsin's "premier food writer" and "the keeper of its culinary heritage." Her columns have run in Edible Madison, Edible Door, Isthmus, Wisconsin Trails and elsewhere, and her books include *The Flavor of Wisconsin*, *The Flavor of Wisconsin for Kids*, *Fresh Market Wisconsin*, *Wisconsin Local Foods Journal* and *The Ovens of Brittany Cookbook*. Terese is a co-founder and longtime leader of the Culinary History Enthusiasts of Wisconsin (CHEW) and past president of the food and sustainability organization, REAP Food Group. She lives in Madison and on Washington Island.